AP Biology 2016
Study Guide

Table of Contents

Quick Overview

As you draw closer to taking your exam, preparing becomes more and more important. Thankfully, you have this study guide to help you get ready. Use this guide to help keep your studying on track and refer to it often.

This study guide contains several key sections that will help you be successful on your exam. The guide contains tips for what you should do the night before and the day of the test. Also included are test-taking tips. Knowing the right information is not always enough. Many well-prepared test takers struggle with exams. These tips will help equip you to accurately read, assess, and answer test questions.

A large part of the guide is devoted to showing you what content to expect on the exam and to helping you better understand that content. Near the end of this guide is a practice test so that you can see how well you have grasped the content. Then, answers explanations are provided so that you can understand why you missed certain questions.

Don't try to cram the night before you take your exam. This is not a wise strategy for a few reasons. First, your retention of the information will be low. Your time would be better used by reviewing information you already know rather than trying to learn lots of new information. Second, you will likely become stressed as you try to gain large amount of knowledge in a short amount of time. Third, you will be depriving yourself of sleep. So be sure to go to bed at a reasonable time the night before. Being well-rested helps you focus and remain calm.

Be sure to eat a substantial breakfast the morning of the exam. If you are taking the exam in the afternoon, be sure to have a good lunch as well. Being hungry is distracting and can make it difficult to focus. You have hopefully spent lots of time preparing for the exam. Don't let an empty stomach get in the way of success!

When travelling to the testing center, leave earlier than needed. That way, you have a buffer in case you experience any delays. This will help you remain calm and will keep you from missing your appointment time at the testing center.

Be sure to pace yourself during the exam. Don't try to rush through the exam. There is no need to risk performing poorly on the exam just so you can leave the testing center early. Allow yourself to use all of the allotted time if needed.

Remain positive while taking the exam even if you feel like you are performing poorly. Thinking about the content you should have mastered will not help you perform better on the exam.

Once the exam is complete, take some time to relax. Even if you feel that you need to take the exam again, you will be well served by some down time before you begin studying again. It's often easier to convince yourself to study if you know that it will come with a reward!

Test-Taking Strategies

1. Predicting the Answer

When you feel confident in your preparation for a multiple-choice test, try predicting the answer before reading the answer choices. This is especially useful on questions that test objective factual knowledge or that ask you to fill in a blank. By predicting the answer before reading the available choices, you eliminate the possibility that you will be distracted or led astray by an incorrect answer choice. You will feel much more confident in your selection if you read the question, predict the answer, and then find your prediction among the answer choices. After using this strategy, be sure to still read all of the answer choices carefully and completely. If you feel unprepared, you should not attempt to predict the answers. This would be a waste of time and an opportunity for your mind to wander in the wrong direction.

2. Reading the Whole Question

Too often, test takers scan a multiple-choice question, recognize a few familiar words, and immediately jump to the answer choices. Test authors are aware of this common impatience, and they will sometimes prey upon it. For instance, a test author might subtly turn the question into a negative, or he or she might redirect the focus of the question right at the end. The only way to avoid falling into these traps is to read the entirety of the question carefully before reading the answer choices.

3. Looking for Wrong Answers

Long and complicated multiple-choice questions can be intimidating. One way to simplify a difficult multiple-choice question is to eliminate all of the answer choices that are clearly wrong. In most sets of answers, there will be at least one selection that can be dismissed right away. If the test is administered on paper, the test taker could draw a line through it to indicate that it may be ignored; otherwise, the test taker will have to perform this operation mentally or on scratch paper. In either case, once the obviously incorrect answers have been eliminated, the remaining choices may be considered. Sometimes identifying the clearly wrong answers will give the test taker some information about the correct answer. For instance, if one of the remaining answer choices is a direct opposite of one of the eliminated answer choices, it may well be the correct answer. The opposite of obviously wrong is obviously right! Of course, this is not always the case. Some answers are obviously incorrect simply because they are irrelevant to the question being asked. Still, identifying and eliminating some incorrect answer choices is a good way to simplify a multiple-choice question.

4. Don't Overanalyze

Anxious test takers often overanalyze questions. When you are nervous, your brain will often run wild causing you to make associations and discover clues that don't actually exist. If you feel that this may be a problem for you, do whatever you can to slow down during the test. Try taking a deep breath or counting to ten. As you read and consider the question, restrict yourself to the particular words used by the author. Avoid thought tangents about what the author *really* meant, or what he or she was *trying* to say. The only things that matter on a multiple-choice test are the words that are actually in the question. You must avoid reading too much into a multiple-choice question, or supposing that the writer meant something other than what he or she wrote.

5. No Need for Panic

It is wise to learn as many strategies as possible before taking a multiple-choice test, but it is likely that you will come across a few questions for which you simply don't know the answer. In this situation, avoid panicking. Because most multiple-choice tests include dozens of questions, the relative value of a single wrong answer is small. Moreover, your failure on one question has no effect on your success elsewhere on the test. As much as possible, you should compartmentalize each question on a multiple-choice test. In other words, you should not allow your feelings about one question to affect your success on the others. When you find a question that you either don't understand or don't know how to answer, just take a deep breath and do your best. Read the entire question slowly and carefully. Try rephrasing the question a couple of different ways. Then, read all of the answer choices carefully. After eliminating obviously wrong answers, make a selection and move on to the next question.

6. Confusing Answer Choices

When working on a difficult multiple-choice question, there may be a tendency to focus on the answer choices that are the easiest to understand. Many people, whether consciously or not, gravitate to the answer choices that require the least concentration, knowledge, and memory. This is a mistake. When you come across an answer choice that is confusing, you need to give it extra attention. A question might be confusing because you do not know the subject matter to which it refers. If this is the case, don't eliminate the answer before you have affirmatively settled on another. When you come across an answer choice of this type, set it aside as you look at the remaining choices. If you can confidently assert that one of the other choices is correct, you can leave the confusing answer aside. Otherwise, you will need to take a moment to try to better understand the confusing answer choice. Rephrasing is one way to tease out the sense of a confusing answer choice.

7. Your First Instinct

Many people struggle with multiple-choice tests because they overthink the questions. If you have studied sufficiently for the test, you should be prepared to trust your first instinct once you have carefully and completely read the question and all of the answer choices. There is a great deal of research to suggest that the mind can come to the correct conclusion very quickly once it has obtained all of the relevant information. At times, it may seem to you as if your intuition is working faster even than your reasoning mind. This may in fact be true. The knowledge you obtain while studying may be retrieved from your subconscious before you have a chance to work out the associations that support it. Verify your instinct by working out the reasons that it should be trusted.

8. Key Words

Many test takers struggle with multiple-choice questions because they have poor reading comprehension skills. Quickly reading and understanding a multiple-choice question requires a mixture of skill and experience. To help with this, try jotting down a few key words and phrases on a piece of scrap paper. Doing this concentrates the process of reading and forces the mind to weigh the relative importance of the question's parts. In selecting words and phrases to write down, the test taker thinks about the question more deeply and carefully. This is especially true for multiple-choice questions that are preceded by a long prompt.

9. Subtle Negatives

One of the oldest tricks in the multiple-choice test writer's book is to subtly reverse the meaning of a question with a word like *not* or *except*. If you are not paying attention to each word in the question, you can easily be led astray by this trick. For instance, a common question format is, "Which of the following is…?" Obviously, if the question instead is, "Which of the following is not….?," then the answer will be quite different. Even worse, the test makers are aware of the potential for this mistake and will include one answer choice that would be correct if the question were not negated or reversed. A test taker who misses the reversal will find what he or she believes to be a correct answer and will be so confident that he or she will fail to reread the question and discover the original error. The only way to avoid this is to practice a wide variety of multiple-choice questions and to pay close attention to each and every word.

10. Reading Every Answer Choice

It may seem obvious, but you should always read every one of the answer choices! Too many test takers fall into the habit of scanning the question and assuming that they understand the question because they recognize a few key words. From there, they pick the first answer choice that answers the question they believe they have read. Test takers who read all of the answer choices might discover that one of the latter answer choices is actually *more* correct. Moreover, reading all of the answer choices can remind you of facts related to the question that can help you arrive at the correct answer. Sometimes, a misstatement or incorrect detail in one of the latter answer choices will trigger your memory of the subject and will enable you to find the right answer. Failing to read all of the answer choices is like not reading all of the items on a restaurant menu. You might miss out on the perfect choice.

11. Spot the Hedges

One of the keys to success on multiple-choice tests is paying close attention to every word. This is never more true than with words like *almost, most, some,* and *sometimes*. These words are called "hedges", because they indicate that a statement is not totally true or not true in every place and time. An absolute statement will contain no hedges, but in many subjects, like literature and history, the answers are not always straightforward. There are always exceptions to the rules in these subjects. For this reason, you should favor those multiple-choice questions that contain hedging language. The presence of qualifying words indicates that the author is taking special care with his or her words, which is certainly important when composing the right answer. After all, there are many ways to be wrong, but there is only one way to be right! For this reason, it is wise when taking a multiple-choice test to avoid answers that are absolute. An absolute answer is one that says things are either all one way or all another. They often include words like *every, always, best,* and *never*. If you are taking a multiple-choice test in a subject that doesn't lend itself to absolute answers, be on your guard if you see any of these words.

12. Long Answers

In many subject areas, the answers are not simple. As already mentioned, the right answer often requires hedges. Another common feature of the answers to a complex or subjective question are qualifying clauses, which are groups of words that subtly modify the meaning of the sentence. If the question or answer choice describes a rule to which there are exceptions or the subject matter is complicated, ambiguous, or confusing, the correct answer will require many words in order to be expressed clearly and accurately. In essence, you should not be deterred by answer choices that seem excessively long. Oftentimes, the author of the text will not be able to write the correct answer without offering some qualifications and modifications. As a test taker, your job is to read the answer choices thoroughly and completely and to select the one that most accurately and precisely answers the question.

13. Restating to Understand

Sometimes, a question on a multiple-choice test is difficult not because of what it asks but because of how it is written. If this is the case, restate the question or answer choice in different words. This process serves a couple of important purposes. First, it forces you to concentrate on the core of the question. In order to rephrase the question accurately, you have to understand it well. Rephrasing the question will concentrate your mind on the key words and ideas. Second, it will present the information to your mind in a fresh way. This process may trigger your memory of some useful scrap of information picked up while studying.

14. True Statements

Sometimes an answer choice will be true in itself, but it does not answer the question. This is one of the main reasons why it is essential to read the question carefully and completely before proceeding to the answer choices. Too often, test takers skip ahead to the answer choices and look for true statements. Having found one of these, they are content to select it without reference to the question above. Obviously, this provides an easy way for test makers to play tricks. The savvy test taker will always read the entire question before turning to the answer choices. Then, having settled on a correct answer choice, he or she will refer to the original question and ensure that the selected answer is relevant. The mistake of choosing a correct-but-irrelevant answer choice is especially common on questions related to specific pieces of objective knowledge, like historical or scientific facts. A prepared test taker will have a wealth of factual knowledge at his or her disposal, but may be careless in its application.

15. No Patterns

One of the more dangerous ideas that circulate about multiple-choice tests is that the correct answers tend to fall into patterns. These erroneous ideas range from a belief that B and C are the most common right answers, to the idea that an unprepared test-taker should answer "A-B-A-C-A-D-A-B-A." It cannot be emphasized enough that pattern-seeking of this type is exactly the WRONG way to approach a multiple-choice test. To begin with, it is highly unlikely that the test maker will plot the correct answers according to some predetermined pattern. The questions are scrambled and delivered in a random order. Furthermore, even if the test maker was following a pattern in the assignation of correct answers, there is no reason why the test maker would know which pattern he or she was using. Any attempt to discern a pattern in the answer choices is a waste of time and a distraction from the real work of taking the test. A test taker would be much better served by extra preparation before the test than by reliance on a pattern in the answers.

Cellular Processes: Energy and Communication

Cells and structural organization

All organisms, whether plants, animals, fungi, protists, or bacteria, exhibit structural organization on the cellular and organism level. All cells contain DNA and RNA and can synthesize proteins. Cells are the basic structural units of all organisms. All organisms have a highly organized cellular structure. Each cell consists of nucleic acids, cytoplasm, and a cell membrane. Specialized organelles such as mitochondria and chloroplasts have specific functions within the cell. In single-celled organisms, that single cell contains all of the components necessary for life. In multicellular organisms, cells can become specialized. Different types of cells can have different functions. Life begins as a single cell whether by asexual or sexual reproduction. Cells are grouped together in tissues. Tissues are grouped together in organs. Organs are grouped together in systems. An organism is a complete individual.

Cellular energy

All cells must obtain and use energy in order to grow, make repairs, and reproduce. Cells use energy to take in food, process that food, and eliminate wastes from this process. Cells obtain the energy they need by the breaking of bonds of molecules. The energy is stored in the chemical bonds of the nutrient molecules. This process of converting this stored energy into usable adenosine triphosphate (ATP) is known as cellular respiration. Organisms differ in how they obtain food. Plants and other autotrophs produce energy through photosynthesis or chemosynthesis. Animals and other heterotrophs obtain their energy from consuming autotrophs or other heterotrophs.

Growth and reproduction

All organisms must be capable of growth and reproduction. Growth is necessary for each organism individually for multicellular organisms to develop and mature and to increase in size. Growth allows cells to be replaced or repaired. All cells eventually die. Without growth from cell division, tissues could not be maintained or repaired. Through mitosis, most cells routinely replace themselves with identical daughter cells. All organisms eventually die. Reproduction is necessary to increase the number of individuals in a population. Reproduction is either sexual by the joining of gametes or asexual by binary fission or some other related method. Not all organisms reproduce, but all must grow or they will die. Even single-celled organisms grow a small amount.

Regulation and responses to the environment

Organisms must be able to adapt to their environment in order to thrive or survive. Individual organisms must be able to recognize stimuli in their surroundings and adapt quickly. For example, an individual euglena can sense light and respond by moving toward the light. Individual organisms must also be able to adapt to changes in the environment on a larger scale. For example, plants must be able to respond to the change in the length of the day to flower at the correct time. Populations must also be able to adapt to a changing environment. Evolution by natural selection is the process by which populations change over many generations. For example, wooly mammoths were unable to adapt to a warming climate and are now extinct, but many species of deer did adapt and are abundant today.

Kingdom systems

In 1735 Carolus Linnaeus devised a two-kingdom classification system. He placed all living things into either the *Animalia* kingdom or the *Plantae* kingdom. Fungi and algae were classified as plants. Also, Linnaeus developed the binomial nomenclature system that is still used today. In 1866, Ernst Haeckel introduced a three-kingdom classification system, adding the *Protista* kingdom to Linnaeus's animal and plant kingdoms. Bacteria were classified as protists. Cyanobacteria were still classified as plants. In 1938, Herbert Copeland introduced a four-kingdom classification system in which bacteria and cyanobacteria were moved to the *Monera* kingdom. In 1969, Robert Whittaker introduced a five-kingdom system that moved fungi from the plant kingdom to the *Fungi* kingdom. Some algae were still classified as plants. In 1977, Carl Woese introduced a six-kingdom system in which in the *Monera* kingdom was replaced with the *Eubacteria* kingdom and the *Archaebacteria* kingdom.

Domain classification system

In 1990, Carl Woese introduced his domain classification system. Domains are broader groupings above the kingdom level. This system consists of three domains- *Archaea*, *Bacteria*, and *Eukarya*. All eukaryotes such as plants, animals, fungi, and protists are classified in the *Eukarya* domain. The *Bacteria* and *Archaea* domains consist of prokaryotes. Organisms previously classified in the *Monera* kingdom are now classified into either the *Bacteria* or *Archaea* domain based on their ribosomal RNA structure. Members of the *Archaea* domain often live in extremely harsh environments.

Viruses

Viruses are nonliving, infectious particles that act as parasites in living organisms. Viruses are acellular, which means that they lack cell structure. Viruses cannot reproduce outside of living cells. The structure of a virus is a core of a nucleic acid, which may be either DNA or RNA, surrounded by a protein coat or capsid. In some viruses, the capsid may be surrounded by a lipid membrane or envelope. Viruses can contain up to 500 genes. Viruses have various shapes and usually are too small to be seen without the aid of an electron microscope. Viruses can infect plants, animals, fungi, protists, and bacteria. Viruses can attack only specific types of cells that have specific receptors on their surfaces. Viruses do not divide or reproduce like living cells. Viruses are replicated by the machinery of the host cell. The nucleic acid of the virus takes control of the host cell's metabolic pathways to make copies of itself. The host cell usually bursts to release these copies.

Bacteria

Bacteria are small, prokaryotic single-celled organisms. Bacteria have a single circular loop of DNA and cytoplasm with ribosomes enclosed in a plasma membrane. A cell wall containing peptidoglycan surrounds the plasma membrane. Some bacteria such as pathogens are further encased in gel-like capsules. Bacteria can be autotrophs such as photosynthetic bacteria found in algae and the chemosynthetic autotrophs found near deep-sea vents or heterotrophs. Some bacteria heterotrophs are saprophytes that function as decomposers in ecosystems, and some are pathogens. Many types of bacteria share commensal or mutualistic relationships with other organisms. Most bacteria reproduce asexually by binary fission. Two identical daughter cells are produced from one parent cell. Some bacteria can transfer genetic material to other bacteria through a process called conjugation. Some bacteria can incorporate DNA from the environment in a process called transformation.

Protists

Protists are small, eukaryotic single-celled organisms. Although protists are small, they are much larger than prokaryotic bacteria. Protists have three general forms, which include plantlike protists, animal-like protists, and funguslike protists. Plantlike protists are algae that contain chlorophyll and perform photosynthesis. Animal-like protists are protozoa with no cell walls that typically lack chlorophyll and are grouped by their method of locomotion. Funguslike protists, which do not have chitin in their cell walls, are generally grouped as either slime molds or water molds. Protists may be autotrophic or heterotrophic. Autotrophic protists include many species of algae. Heterotrophic protists include parasitic, commensalistic, and mutualistic protozoa. Slime molds are heterotrophic funguslike protists, which consume microorganisms. Some protists reproduce sexually, but most reproduce asexually by binary fission. Some reproduce asexually by spores. Some reproduce by alternation of generations and require two hosts in their life cycle.

Fungi

Fungi are nonmotile organisms with eukaryotic cells containing chitin in their cell walls. Most fungi are multicellular, but a few including yeast are unicellular. Fungi have multicellular filaments called hyphae that are grouped together in mycelia. Fungi do not perform photosynthesis. All fungi are heterotrophs. Fungi can be parasitic, mutualistic or free living. Free-living fungi include mushrooms and toadstools. Parasitic fungi include fungi responsible for ringworm and athlete's foot. Mycorrhizae are mutualistic fungi that live in or near plant roots increasing the roots' surface area of absorption. Almost all fungi reproduce asexually by spores, but most fungi also have a sexual phase in the production of spores. Some fungi reproduce by budding or fragmentation.

Plants

Plants are multicellular organisms with eukaryotic cells containing cellulose in their cell walls. Plant cells have chlorophyll and perform photosynthesis. Plants can be vascular or nonvascular. Vascular plants have true leaves, stems, and roots that contain xylem and phloem. Nonvascular plants lack true leaves, stems and roots and do not have any true vascular tissue but instead rely on diffusion and osmosis for most transport of materials. Almost all plants are autotrophic, relying on photosynthesis for food. A small number do not have chlorophyll and are parasitic, but these are extremely rare. Plants can reproduce sexually or asexually. Many plants reproduce by seeds produced in the fruits of the plants. Some plants reproduce by seeds on cones. Ferns reproduce by spores. Some plants can reproduce asexually by vegetative reproduction.

Animals

Animals are multicellular organism with eukaryotic cells that do not have cell walls surrounding their plasma membranes. Animals have several possible structural body forms. Animals can be relatively simple in structure such as sponges, which do not have a nervous system. Other animals are more complex with cells organized into tissues, and tissues organized into organs, and organs even further organized into systems. Invertebrates such as arthropods, nematodes, and annelids have complex body systems. Vertebrates including fish, amphibians, reptiles, birds, and mammals are the most complex with detailed systems such as those with gills, air sacs, or lungs designed to exchange respiratory gases. All animals are heterotrophs and obtain their nutrition by consuming autotrophs or other heterotrophs. Most animals are motile, but some animals move their environment to bring food to them. All animals reproduce sexually at some point in their life cycle. Typically, this involves the union of a sperm and egg to produce a zygote.

Body planes

Animals can exhibit bilateral symmetry, radial symmetry, or asymmetry. With bilateral symmetry, the organism can be cut in half along only one plane to produce two identical halves. Most animals, including all vertebrates such as mammals, birds, reptiles, amphibians, and fish, exhibit bilateral symmetry. Many invertebrates including arthropods and crustaceans also exhibit bilateral symmetry. With radial symmetry, the organism can be cut in half along several planes to produce two identical halves. Starfish, sea urchins, and jellyfish exhibit radial symmetry. With asymmetry, the organism exhibits no symmetry. Very few organisms in the animal phyla exhibit asymmetry, but a few species of sponges are asymmetrical.

Body cavities

Animals can be grouped based on their types of body cavities. A coelom is a fluid-filled body cavity between the alimentary canal and the body wall. The three body plans based on the formation of the coelom are acoelomates, pseudocoelomates, and coelomates. Acoelomates do not have body cavities. Pseudocoelomates have a body cavity called a pseudocoelom. Pseudocoeloms are not considered true coeloms. Pseudocoeloms are located between mesoderm and endoderm instead of actually in the mesoderm as in a true coelom. Coelomates have a true coelom located within the mesoderm. Simple or primitive animals such as sponges, jellyfish, sea anemones, hydras, flatworms, and ribbon worms are acoelomates. Pseudocoelomates include roundworms and rotifers. Most animals including arthropods, mollusks, annelids, echinoderms, and chordates are coelomates.

Modes of reproduction

Animals can reproduce sexually or asexually. Most animals reproduce sexually. In sexual reproduction, males and females have different reproductive organs that produce gametes. Males have testes that produce sperm, and females have ovaries that produce eggs. During fertilization, a sperm cell unites with an egg cell, forming a zygote. Fertilization can occur internally such as in most mammals and birds or externally such as aquatic animals such as fish and frogs. The zygote undergoes cell division, which develops into an embryo and eventually develops into an adult organism. Some embryos develop in eggs such as in most fish, amphibians, reptiles, and birds. Some mammals are oviparous and lay eggs. Most mammals are viviparous and have a uterus in which the embryo develops. Some mammals are marsupials and give birth to an immature fetus that finishes developing in a pouch. Some animals reproduce asexually. For example, hydras reproduce by budding, and starfish and planarians can reproduce by fragmentation and regeneration. Some fish, frogs, and insects reproduce by parthenogenesis.

Temperature regulation

Animals can be classified as either homeotherms or poikilotherms. Homeotherms, also called warm-blooded animals or endotherms, maintain a constant body temperature regardless of the temperature of the environment. Homeotherms such as mammals and birds have a high metabolic rate because much energy is needed to maintain the constant temperature. Poikilotherms, also called cold-blooded animals or ectotherms, do not maintain a constant body temperature. Their body temperature fluctuates with the temperature of the environment. Poikilotherms such as arthropods, fish, amphibians, and reptiles have metabolic rates that fluctuate with their body temperature.

Cells

Cells are the basic structural units of all living things. Cells are composed of various molecules including proteins, carbohydrates, lipids, and nucleic acids. All animal cells are eukaryotic. All animal cells have a nucleus, cytoplasm, and a cell membrane. Organelles include mitochondria, ribosomes, endoplasmic reticulum, Golgi apparatuses, and vacuoles. Specialized cells are numerous including but not limited to various muscle cells, nerve cells, epithelial cells, bone cells, blood cells, and cartilage cells. Cells are grouped to together in tissues to perform specific functions.

Organizational hierarchy of multicellular organisms

Multicellular organisms are made up of cells, which are grouped together in tissues. Tissues are grouped together in organs. Organs are grouped together into organ systems. Organs systems are grouped together into a single organism. Cells are defined as the basic structural units of an organism. Cells are the smallest living units. Tissues are groups of cells that work together to perform a specific function. Organs are groups of tissues that work together to perform a specific function. Organ systems are groups of organs that work together to perform a specific function. An organism is an individual that contains several body systems.

Tissues

Tissues are groups of cells that work together to perform a specific function. Tissues can be grouped into four broad categories: muscle tissue, nerve tissue, epithelial tissue, and connective tissue. Muscle tissue is involved in body movement. Muscle tissues can be composed of skeletal muscle cells, cardiac muscle cells, or smooth muscle cells. Skeletal muscles include the muscles commonly called biceps, triceps, hamstrings, and quadriceps. Cardiac muscle tissue is found only in the heart. Smooth muscle tissue provides tension in the blood vessels, control pupil dilation, and aid in peristalsis. Nerve tissue is located in the brain, spinal cord, and nerves. Epithelial tissue makes up the layers of the skin and various membranes. Connective tissues include bone tissue, cartilage, tendons, ligaments, fat, blood, and lymph. Tissues are grouped together as organs to form specific functions.

Organs

Organs are groups of tissues that work together to perform specific functions. Complex animals have several organs that are grouped together in multiple systems. For example, the heart is specifically designed to pump blood throughout an organism's body. The heart is composed mostly of muscle tissue in the myocardium, but it also contains connective tissue in the blood and membranes, nervous tissue that controls the heart rate, and epithelial tissue in the membranes. Gills in fish and lungs in reptiles, birds, and mammals are specifically designed to exchange gases. In birds, crops are designed to store food and gizzards are designed to grind food.

Organ systems are groups of organs that work together to perform specific functions. In mammals, there are 11 major organ systems: integumentary system, respiratory system, cardiovascular system, endocrine system, nervous system, immune system, digestive system, excretory system, muscular system, skeletal system, and reproductive system. For example, in mammals, the cardiovascular system that transports materials throughout the body consists of the heart, blood vessels, and blood. The respiratory system, which provides for the exchange of gases, consists of the nasal passages, pharynx, larynx, trachea, bronchial tubes, lungs, alveoli, and diaphragm. The digestive system, which processes consumed food, consists of the alimentary canal and additional organs including the liver, gallbladder, and pancreas.

Cardiovascular system

The cardiovascular system consists primarily of the heart, blood, and blood vessels. The heart is a pump that pushes blood through the arteries. Arteries are blood vessels that carry blood away from the heart, and veins are blood vessels that carry blood back to the heart. The exchange of materials between blood and cells occur in the capillaries, which are the tiniest of the blood vessels. Blood is the fluid that carries materials to and from each cell of an organism. The main function of the cardiovascular system is to provide for gas exchange, the delivery of nutrients and hormones, and waste removal. All vertebrates and a few invertebrates including annelids, squids, and octopuses have a closed circulatory system. Blood is pumped through a series of vessels and does not fill body cavities. Mammals, birds and crocodilians have a four-chambered heart. Most amphibians and reptiles have a three-chambered heart. Fish have only a two-chambered heart. Arthropods and most mollusks have open circulatory systems. Usually blood is pumped by a heart into the body cavities and bathes the tissues in blood. Muscle movement moves the blood through the body. Blood then diffuses back to the heart through the cells. Many invertebrates do not have a cardiovascular system. For example, echinoderms have a water vascular system.

Respiratory system

The function of the respiratory system is to move air in and out of the body in order to facilitate the exchange of oxygen and carbon dioxide. The respiratory system consists of the nasal passages, pharynx, larynx, trachea, bronchial tubes, lungs, and diaphragm. When the diaphragm contracts, the volume of the chest increases, which reduces the pressure in the lungs. The intercostal muscles also aid in breathing. Then, air is inhaled through the nose or mouth and passes through the pharynx, larynx, trachea, and bronchial tubes into the lungs. Bronchial tubes branch into bronchioles, which end in clusters of alveoli. The alveoli are tiny sacs inside the lungs where gas exchange takes place. When the diaphragm relaxes, the volume in the chest cavity decreases, forcing the air out of the lungs.

Reproductive system

The main function of the reproductive system is to propagate the species. Most animals reproduce sexually at some point in their life cycle. Typically, this involves the union of a sperm and egg to produce a zygote. Eggs may be fertilized internally or externally. In complex animals, the female reproductive system includes an ovary, which produces the egg cell. The male reproductive system includes a teste, which produces the sperm. In internal fertilization in mammals, the sperm unites with the egg in the oviduct. In mammals, the zygote begins to divide, and the blastula implants in the uterus. Most mammals give birth to live young, but monotremes lay eggs. In birds, after the egg is fertilized, albumen, membranes, and egg shell are added. Reptiles lay amniotic eggs covered by a leathery shell. Amphibians lay eggs in water or moist areas, and the eggs are fertilized externally. Most fish lay eggs in water to be fertilized externally, but a few fish give birth to live young. Most invertebrates reproduce sexually. Invertebrates may have separate sexes or be hermaphroditic, in which the organisms produces sperm and eggs either at the same time or separately at some time in their life cycle. Many invertebrates such as insects also have complex reproductive systems. Some invertebrates reproduce asexually by budding, fragmentation, or parthenogenesis.

Digestive system

The main function of the digestive system is to process the food that is consumed by the animal. This includes mechanical and chemical processing. Depending on the animal, mechanical processes can happen in various ways. Mammals have teeth to chew their food. Saliva is secreted, which contains enzymes to begin the breakdown of starches. Many animals such as birds, earthworms, crocodilians, and crustaceans have a gizzard or gizzard-like organ that grinds the food. Many animals such as mammals, birds, reptiles, amphibians, and fish have a stomach that stores and absorbs food. Gastric juice containing enzymes and hydrochloric acid is mixed with the food. The intestine or intestines absorb nutrients and reabsorb water from the undigested material. Many animals have a liver, gallbladder, and pancreas, which aid in digestion of proteins and fats although not being part of the muscular tube through which the waste passes. Undigested wasted are eliminated from the body through an anus or cloaca.

Excretory system

All animals have some type of excretory system that has the main function of processing and eliminating metabolic wastes. In complex animals such as mammals, the excretory system consists of the kidneys, ureters, urinary bladder, and urethra. Urea and other toxic wastes must be eliminated from the body. The kidneys constantly filter the blood. The nephron is the working unit of the kidney. Each nephron functions like a tiny filter. Nephrons not only filter the blood, but they also facilitate reabsorption and secretion. Basically, the glomerulus filters the blood. Water and dissolved materials such as glucose and amino acids pass on into the Bowman's capsule. Depending on concentration gradients, water and dissolved materials can pass back into the blood primarily through the proximal convoluted tubule. Additional water can be removed at the loop of Henle. Antidiuretic hormone regulates the water that is lost or reabsorbed. Urine passes from the kidneys through the ureters to the urinary bladder where it is stored before it is expelled from the body through the urethra.

Nervous system

All animals except sponges have a nervous system. The main function of the nervous system is to coordinate the activities of the body. The nervous system consists of the brain, spinal cord, peripheral nerves, and sense organs. Sense organs such as the ears, eyes, nose, taste buds, and pressure receptors receive stimuli from the environment and relay that information through nerves and the spinal cord to the brain where the information is processed. The brain sends signals through the spinal cord and peripheral nerves to the organs and muscles. The autonomic nervous system controls all routine body functions by the sympathetic and parasympathetic divisions. Reflexes, which are also part of the nervous system, may involve only a few nerve cells and bypass the brain when an immediate response is necessary.

Endocrine system

The endocrine system consists of several ductless glands, which secrete hormones directly into the bloodstream. The pituitary gland is the master gland, which controls the functions of the other glands. The pituitary gland regulates skeletal growth and the development of the reproductive organs. The pineal gland regulates sleep cycles. The thyroid gland regulates metabolism and helps regulate the calcium level in the blood. The parathyroid glands also help regulate the blood calcium level. The adrenal glands secrete the emergency hormone epinephrine, stimulate body repairs, and regulate sodium and potassium levels in the blood. The islets of Langerhans located in the pancreas secrete insulin and glucagon to regulate the blood sugar level. In females, ovaries produce estrogen, which stimulates sexual development, and progesterone, which functions during pregnancy. In males, the testes secrete testosterone, which stimulates sexual development and sperm production.

Immune system

The immune system in animals defends the body against infection and disease. The immune system can be divided into two broad categories: innate immunity and adaptive immunity. Innate immunity includes the skin and mucous membranes, which provide a physical barrier to prevent pathogens from entering the body. Special chemicals including enzymes and proteins in mucus, tears, sweat, and stomach juices destroy pathogens. Numerous white blood cells such as neutrophils and macrophages protect the body from invading pathogens. Adaptive immunity involves the body responding to a specific antigen. Typically, B-lymphocytes or B cells produce antibodies against a specific antigen, and T-lymphocytes or T-cells take special roles as helpers, regulators, or killers. Some T-cells function as memory cells.

Feedback mechanisms

Feedback mechanisms play a major role in the homeostasis in organisms. Each feedback mechanism consists of receptors, an integrator, and effectors. Receptors such as mechanoreceptors or thermoreceptors in the skin detect the stimuli. The integrator such as the brain or spinal cord receives the information concerning the stimuli and sends out signals to other parts of the body. The effectors such as muscles or glands respond to the stimulus. Basically, the receptors receive the stimuli and notify the integrator, which signals the effectors to respond. Feedback mechanisms can be negative or positive. Negative-feedback mechanisms are mechanisms that provide a decrease in response with an increase in stimulus that inhibits the stimulus, which in turn decreases the response. Positive-feedback mechanisms are mechanisms that provide an increase in response with an increase in stimulus, which actually increases the stimulus, which in turn increases the response.

Hypothalamus

The hypothalamus plays a major role in the homoeostasis of vertebrates. Homeostasis is regulation of internal chemistry to maintain a constant internal environment. The hypothalamus is the central portion of the brain just above the brainstem, which is linked to the endocrine system through the pituitary gland. The hypothalamus releases special hormones that influence the secretion of pituitary hormones. The hypothalamus regulates the fundamental physiological state by controlling body temperature, hunger, thirst, sleep, behaviors related to attachment, sexual development, fight-or-flight stress response, and circadian rhythms.

Hormones

All vertebrates have an endocrine system that consists of numerous ductless glands that produce hormones that help coordinate many functions of the body. Hormones are messengers or molecules that transmit signals to other cells. Hormones can consist of amino acids, proteins, or lipid molecules such as steroid hormones. Hormones can affect target cells, which have the correct receptor that is able to bind to that particular hormone. Most cells have receptors for more than one type of hormone. Hormones are distributed to the target cells in the blood by the cardiovascular system. Hormones incorporate feedback mechanisms to help the body maintain homeostasis. Hormones are signaling molecules that are received by receptors. Many hormones are secreted in response to signals from the pituitary gland and hypothalamus gland. Other hormones are secreted in response to signals from inside the body.

Antidiuretic hormone

Antidiuretic hormone (ADH) helps maintain homeostasis in vertebrates. ADH is produced by the posterior pituitary gland, and it regulates the reabsorption of water in the kidneys and concentrates the urine. The stimulus in this feedback mechanism is a drop in blood volume due to water loss. This signal is picked up by the hypothalamus, which signals the pituitary gland to secrete ADH. ADH is carried by the cardiovascular system to the nephrons in the kidneys signaling them to reabsorb more water and send less out as waste. As more water is reabsorbed, the blood volume increases, which is monitored by the hypothalamus. As the blood volume reaches the set point, the hypothalamus signals for a decrease in the secretion of ADH, and the cycle continues.

Insulin and glucagon

Insulin and glucagon are hormones that help maintain the glucose concentration in the blood. Insulin and glucagon are secreted by the clumps of endocrine cells called the pancreatic islets that are located in the pancreas. Insulin and glucagon work together to maintain the blood glucose level. Insulin stimulates cells to remove glucose from the blood. Glucagon stimulates the liver to convert glycogen to glucose. After eating, glucose levels increase in the blood. This stimulus signals the pancreas to stop the secretion of glucagon and to start secreting insulin. Cells respond to the insulin and remove glucose from the blood, lowering the level of glucose in the blood. Later, after eating, the level of glucose in the blood decreases further. This stimulus signals the pancreas to secrete glucagon and decrease the secretion of insulin. In response to the stimulus, the liver converts glycogen to glucose, and the level of glucose in the blood rises. When the individual eats, the cycle begins again.

Behaviors

Animals exhibit many adaptations that help them achieve homeostasis, or a stable internal environment. Some of these adaptions are behavioral. Most organisms exhibit some type of behavioral thermoregulation. Thermoregulation is the ability to keep the body temperature within certain boundaries. The type of behavioral thermoregulation depends on whether the animal is an endotherm or an ectotherm. Ectotherms are "cold-blooded," and their body temperature changes with their external environment. To regulate their temperature, ectotherms often move to an appropriate location. Fish move to warmer waters. Animals will climb to higher grounds. Diurnal ectotherms such as reptiles often bask in the sun to increase their body temperatures. Lizards can alternate between basking and retreating to the shade to maintain a relatively stable body temperature. Butterflies are heliotherms in that they derive nearly all of their heat from basking in the sun. Endotherms are "warm-blooded" and maintain a stable body temperature by internal means. However, many animals that live in hot environments have adapted to the nocturnal lifestyle. Desert animals are often nocturnal to escape high daytime temperatures. Other nocturnal animals sleep during the day in underground burrows or dens. Birds that dive for food in cold waters can spread their wings to capture the heat of the incoming sun.

Kidney

The kidney has a major role in homeostasis of many animals. Homeostasis is regulation of internal chemistry to maintain a constant internal environment. The kidneys filter the blood and remove the nitrogenous wastes. The kidneys also regulate the pH balance throughout the body. First, the blood is filtered. Filtration takes place in the nephron's glomerulus. Next, reabsorption occurs. Reabsorption takes place in the loop of Henle and the conducting duct. The kidneys regulate this reabsorption of water, salt, and other nutrients to maintain the blood volume and proper ion balance. Last, metabolic nitrogenous wastes including urea, ammonia, and uric acid are excreted.

Gamete formation

Gametogenesis is the formation of gametes. Gametes are reproductive cells. Gametes are produced by meiosis. Meiosis is a special type of cell division that consists of two consecutive mitotic divisions referred to as meiosis I and meiosis II. Meiosis I is a reduction division in which a diploid cell is reduced to two haploid daughter cells that contain only one of each pair of homologous chromosomes. During meiosis II, those haploid cells are further divided to form four haploid cells. Spermatogenesis in males produces four viable sperm cells from each complete cycle of meiosis. Oogenesis produces four daughter cells, but only one is a viable egg and the other three are polar bodies.

Fertilization

Fertilization is the union of a sperm cell and an egg cell to produce a zygote. Many sperm may bind to an egg, but only one joins with the egg and injects its nuclei into the egg. Fertilization can be external or internal. External fertilization takes place outside of the female's body. For example, many fish, amphibians, crustaceans, mollusks, and corals reproduce externally by spawning or releasing gametes into the water simultaneously or right after each other. Mammals, birds, reptiles, and some fish rely on internal fertilization. Reptiles and birds reproduce by internal fertilization. All mammals except monotremes reproduce by internal fertilization. Most birds press their cloacae together to transmit the sperm to the egg. Male fish use their anal fin to direct sperm into the female.

Embryonic development

Embryonic development in animals is typically divided into four stages: cleavage, patterning, differentiation, and growth. Cleavage occurs immediately after fertilization when the large single-celled zygote immediately begins to divide into smaller and smaller cells without an increase in mass. A hollow ball of cells forms a blastula. Next, during patterning, gastrulation occurs. During gastrulation, the cells are organized into three primary germ layers: ectoderm, mesoderm, and endoderm. Then, the cells in these layers differentiate into special tissues and organs. For example, the nervous system develops from the ectoderm. The muscular system develops from the mesoderm. Much of the digestive system develops from the endoderm. The final stage of embryonic development is growth and further tissue specialization. The embryo continues to grow until ready for hatching or birth.

Postnatal growth, development, and aging

Postnatal growth occurs from hatching or birth until death. The length of the postnatal growth depends on the species. Elephants can live 70 years, but mice only about 4 years. Chickens can live 15 years. Worker bees can live 1 year. Right after animals are hatched or born, they go through a period of rapid growth and development. In vertebrates, bones lengthen, muscles grow in bulk, and fat is deposited. At maturity, bones stop growing in length, but bones can grow in width and repair themselves throughout the animal's lifetime, and muscle deposition slows down. Fat cells continue to increase and decrease in size throughout the animal's life. Growth is controlled by genetics but is also influenced by nutrition and disease. Most animals are sexually mature in less than two years and can produce offspring.

Vascular and nonvascular plants

Vascular plants, also referred to as tracheophytes, have dermal tissue, meristematic tissue, ground tissues, and vascular tissues. Nonvascular plants, also referred to a bryophytes, do not have the vascular tissue xylem and phloem. Vascular plants can grow very tall, whereas nonvascular plants are short and close to the ground. Vascular plants can be found in dry regions, but nonvascular plants typically grow near or in moist areas. Vascular plants have leaves, roots, and stems, but nonvascular plants have leaflike, rootlike, and stemlike structures that do not have true vascular tissue. Mosses and liverworts, for example, have tiny rootlike structures called rhizoids. Vascular plants include angiosperms, gymnosperms, and ferns. Nonvascular plants include mosses and liverworts.

Flowering and nonflowering plants

Angiosperms and gymnosperms are both vascular plants. Angiosperms are flowering plants, and gymnosperms are nonflowering plants. Angiosperms reproduce by seeds that are enclosed in an ovary, usually in a fruit. Angiosperms can be further classified as either monocots or dicots. Gymnosperms reproduce by unenclosed or "naked" seeds on scales, leaves, or cones. Angiosperms include grasses, weeds, garden flowers, and vegetables and broadleaf trees such as maples, birches, elms, and oaks. Gymnosperms include conifers such as pines, spruces, cedars, and redwoods, cycads, and gingkos.

Monocots and dicots

Angiosperms can be classified as either monocots or dicots. The seeds of monocots have one cotyledon, and the seeds of dicots have two cotyledons. The flowers of monocots have petals in multiples of three, and the flowers of dicots have petals in multiples of four or five. The leaves of monocots are slender with parallel veins, and the leaves of dicots are broad and flat with branching veins. The vascular bundles in monocots are distributed throughout the stem. The vascular bundles in dicots are arranged in rings. Monocots have a fibrous root system, and dicots have a taproot system.

Plant dermal tissue

Plant dermal tissue consists of the epidermis and the dermis. The epidermis is usually a single layer of cells that covers younger plants. The epidermis protects the plant by secreting the cuticle, which is a waxy substance that helps prevent water loss and infections. The epidermis in leaves has tiny pores called stomata. Guard cells in the epidermis control the opening and closing of the stomata. The epidermis usually does not have chloroplasts. The epidermis may be replaced by periderm in older plants. The periderm is also referred to as bark. The layers of the periderm are cork cells or phellem, phelloderm, and cork cambium or phellogen. Cork is the outer layer of the periderm and consists of nonliving cells. The periderm protects the plant and provides insulation.

Vascular tissue

The two major types of plant vascular tissue are xylem and phloem. Xylem is made up of tracheids and vessel elements. All vascular plants contain tracheids, but only angiosperms contain vessel elements. Xylem provides support and conducts water and dissolved minerals from the root and upward throughout the plant by transpirational pull and root pressure. In woody plants, xylem is commonly referred to as wood. Phloem is made up of companion cells and sieve-tube cells. Phloem conducts nutrients including sucrose produced during photosynthesis and organic materials throughout the plant. By active transport, the companion vessels move glucose in and out of the sieve-tube cells. A meristem called vascular cambium is located between the xylem and phloem and produces new xylem and phloem. Xylem and phloem are bound together in vascular bundles.

Ground tissue

The three major types of ground tissue are parenchyma tissue, collenchyma tissue, and sclerenchyma tissue. Most ground tissue is made up of parenchyma. Parenchyma is formed by parenchyma cells, and it provides photosynthesis, food storage, and tissue repair. The soft "filler" tissues in plants are usually parenchyma. The mesophyll in leaves is parenchyma tissue. Collenchyma is made of collenchyma cells and provides support in roots, stems, and petioles. Sclerenchyma tissue is made of sclereid cells, which are more rigid than the collenchyma cells, and provides rigid support and protection. Plant sclerenchyma tissue may contain cellulose or lignin. Fabrics such as jute, hemp, and flax are made of sclerenchyma tissue.

Meristematic tissue

Meristems or meristematic tissues are the regions of plant growth. The cells in meristems are undifferentiated and always remain totipotent, which means they can always develop into any type of special tissue. Meristem cells produce all the new cells in a plant and regenerate damaged parts. Cells of meristems reproduce asexually through mitosis or cell division that is regulated by hormones. The two types of meristems are lateral meristems and apical meristems. Primary growth occurs at apical meristems. Roots and shoots have meristem tissue at their tips and can grow in length. Primary meristems include the protoderm, which produces epidermis; the procambium, which produces cambium; xylem and phloem; and the ground meristem, which produces ground tissue including parenchyma. Secondary growth occurs at the lateral or secondary meristems. Secondary meristems include the vascular cambium and cork cambium. Secondary growth causes an increase in diameter.

Flowers

The primary function of flowers is to produce seeds for reproduction of the plant. Flowers have a pedicel or stalk with a receptacle or enlarged upper portion, which holds the developing seeds. Flowers also can have sepals and petals. Sepals are leaflike structures and protect the bud. Petals, which are collectively called the corolla, help to attract pollinators. Plants can have stamens, pistils, or both depending on the type of plant. The stamen consists of the anther and filament. The anther produces the pollen, which produces the sperm cells. The pistil consists of the stigma, style, and ovary. The ovary contains the ovules, which house the egg cells.

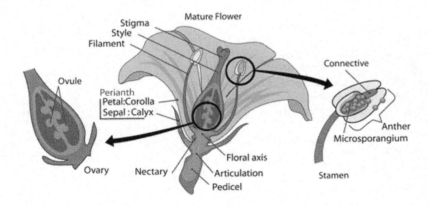

Stems

FPlants can have either woody or nonwoody (herbaceous) stems. The stem is divided into nodes and internodes. Buds are located at the nodes and may develop into leaves, roots, flowers, cones, or more stems. Stems consist of dermal tissue, ground tissue, and vascular tissue. Dicot stems have vascular bundles distributed through the stem. Monocots have rings of vascular bundles. Stems have four main functions: (1) they provide support to leaves, flowers, and fruits; (2) they transport materials in the xylem and phloem; (3) they store food; and (4) they have meristems, which provide all of the new cells for the plant.

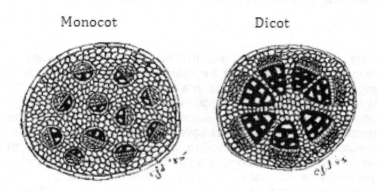

Monocot   Dicot

Leaf

The primary function of a leaf is to manufacture food through photosynthesis. The leaf consists of a flat portion called the blade and a stalk called the petiole. The edge of the leaf is called the margin and can be entire, toothed, or lobed. Veins transport food and water and make up the skeleton of the leaf. The large central vein is called the midrib. The blade has an upper and lower epidermis. The epidermis is covered by a protective cuticle. The lower epidermis contains many stomata or pores, which allow air to enter and leave the leaf. Stomata also regulate transpiration. The middle portion of the leaf is called the mesophyll. The mesophyll consists of the palisade mesophyll and the spongy mesophyll. Most photosynthesis occurs in chloroplasts located in the palisade mesophyll.

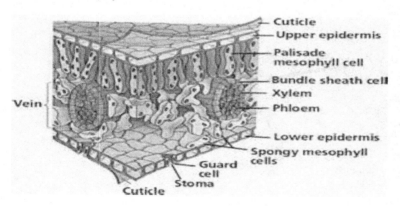

20

Roots

The primary functions of roots are to anchor the plant, absorb water and materials, and store food. The two basic types of root systems are taproot systems and fibrous root systems. Taproot systems have a primary root with many smaller secondary roots. Fibrous root systems, which lack a primary root, consist of a mass of many small secondary roots. The root has three main regions: the area of maturation, the area of elongation, and the area of cell division or the meristematic region. The root is covered by an epidermal cell, some of which develops into root hairs. Root hairs absorb water and minerals by osmosis, and capillarity helps move the water upward through the roots to the rest of the plant. The center of the root is the vascular cylinder, which contains the xylem and phloem. The vascular cylinder is surrounded by the cortex where the food is stored. Primary growth occurs at the root tip. Secondary growth occurs at the vascular cambium located between the xylem and phloem.

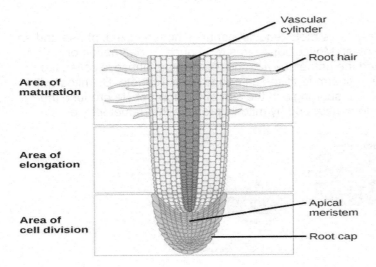

Pollination strategies

Pollination is the transfer of pollen from the anther of the stamen to the stigma of the pistil on the same plant or on a different plant. Pollinators can be either abiotic or biotic. Abiotic pollinators include wind and water. Approximately 20% of pollination occurs by abiotic pollinators. For example, grasses are typically pollinated by wind, and aquatic plants are typically pollinated by water. Biotic pollinators include insects, birds, mammals, and occasionally reptiles. Most biotic pollinators are insects. Many plants have colored petals and strong scents, which attract insects. Pollen rubs off on the insects and is transferred as they move from plant to plant.

Seed dispersal

Methods of seed dispersal can be abiotic or biotic. Methods of seed dispersal include gravity, wind, water, and animals. Some plants produce seeds in fruits that get eaten by animals and then are distributed to new locations in the animals' waste. Some seeds have wings or featherlike structures to aid in dispersal by wind. For example, dandelion seeds are dispersed by the wind. Some seeds have barbs that get caught in animal hair or bird feathers and are then carried to new locations by the animals. Some animals bury seeds for food storage but do not return for the seeds. The seeds of aquatic plants can be dispersed by water. The seeds of plants near rivers, streams, lakes, and beaches are often dispersed by water. For example, coconuts are often dispersed by water. Some plants, in a method called mechanical dispersal, can propel or shoot their seeds away from them even up to several feet. For example, touch-me-nots and violets can reproduce by mechanical dispersal.

Alternation of generations

Alternation of generations, also referred to as metagenesis, contains both a sexual phase and an asexual phase in the life cycle of the plant. Mosses and ferns reproduce by alternation of generations: the sexual phase is called the gametophyte, and the asexual phase is called the sporophyte. During the sexual phase, a sperm fertilizes an egg to form a zygote. By mitosis, the zygote develops into the sporophyte. The sporangia in the sori of the sporophyte produce the spores through meiosis. The spores germinate and by mitosis produce the gametophyte.

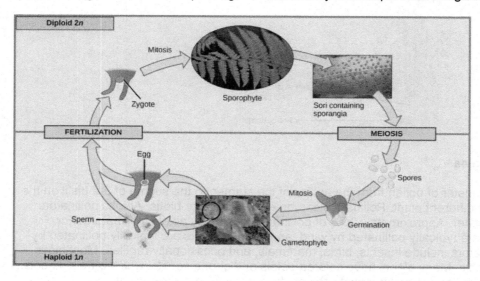

Obtaining and transporting water and inorganic nutrients

Water and inorganic nutrients enter through the plant's root hairs by osmosis and active transport and pass through the endodermis to the xylem. More than 90% of a plant's water is lost through the stomata in the epidermis of the plant by transpiration. This loss is necessary to provide the tension needed to pull the water and nutrients up through the xylem. In order to maintain the remaining water that is necessary for the functioning of the plant, guard cells control the stomata. Whether an individual stoma is closed or open is controlled by two guard cells. When the guard cells lose water and become flaccid, they collapse together, closing the stoma. When the guard cells swell with water and become turgid, they move apart, opening the stoma.

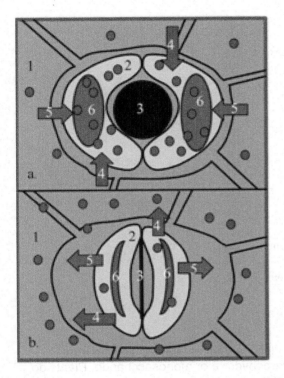

Water and inorganic nutrients enter plants through the root hair and travel to the xylem. Once the water, minerals, and salts have crossed the endodermis, they must be moved upward through the xylem by water uptake. Most of a plant's water is lost through the stomata by transpiration. This is necessary for water uptake to occur. The xylem contains dead water-conducting cells called tracheids and vessels. The movement of water upward through the tracheids and vessels is explained by the cohesion-tension theory. First, water is lost through evaporation of the plant's surface through transpiration. This can occur at any surface exposed to air but is mainly through the stomata in the epidermis. This transpiration puts the water inside the xylem in a state of tension. Because water is cohesive due to the strong hydrogen bonds between molecules, the water is pulled up the xylem as long as the water is transpiring.

Plant roots have numerous root hairs that absorb water and inorganic nutrients such as minerals and salts. Root hairs are thin, hairlike outgrowths of the root's epidermal cells that exponentially increase the root's surface area. Water molecules cross the cell membranes of the root hairs by osmosis and then travel on to the vascular cylinder. Inorganic nutrients are transported across the cell membranes of the root endodermis by active transport. The endodermis is a single layer of cells that the water and nutrients must pass through by osmosis or active transport. Casparian strips, which are waxy waterproof deposits, line the channels between the cells of the endodermis to prevent crossing there. Water passes through by osmosis, but mineral uptake is controlled by transport proteins.

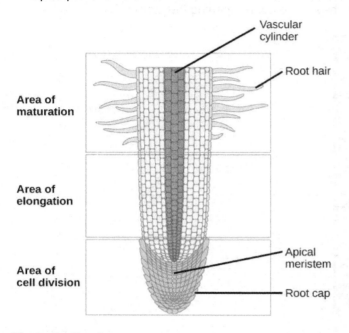

Photosynthesis

Plants produce glucose during photosynthesis. That glucose then enters reactions to form sucrose, starch, and cellulose. Glucose is a simple carbohydrate or monosaccharide. Plants do not transport glucose molecules. Instead, the glucose is joined to a fructose to form a sucrose, which is transported in sap. Sucrose is a disaccharide. Glucose and sucrose are simple carbohydrates. Starches and cellulose are long chains of glucose molecules called polysaccharides. Plants store glucose as starch, and plants use cellulose for rigidity in their cell walls. Both starch and cellulose are complex carbohydrates.

The movement of sugars and other materials from the leaves to other tissues throughout the plants is called translocation. Nutrients are translocated from sources (areas with excess sugars) such as mature leaves to sinks (areas where sugars are needed) such as flowers, fruits, developing leaves, and roots. Phloem vessels are found in the vascular bundles along with the xylem. Phloem contains conducting cells called sieve elements, which are connected end to end in sieve tubes. Sieve tubes carry sap from sugar sources to sugar sinks. Phloem sap contains mostly sucrose dissolved in water. The sap can also contain proteins, amino acids, and hormones. Some plants transport sugar alcohols. Loading the sugar into the sieve tubes causes water to enter the tubes by osmosis, creating a higher hydrostatic pressure at the source end of the tube. Sugar is removed from the sieve tube at the sink end, and water again follows by osmosis lowering the pressure. This process is referred to as the pressure-flow mechanism.

Plastids are major organelles found in plants and algae. Because plastids can differentiate, there are many forms of plastids. For example, chloroplasts are plastids. Specialized plastids can store pigments, starches, fats, or proteins. Amyloplasts are the plastids that store the starch formed from long chains of glucose produced during photosynthesis. Amyloplasts synthesize and store the starch granules through the polymerization of glucose. When needed, amyloplasts also convert these starch granules back into sugar. Fruits and potato tubers have large numbers of amyloplasts. Chloroplasts can also synthesize and store starch. Interestingly, amyloplasts can redifferentiate and transform into chloroplasts.

Atomic structure of atoms

Important biological molecules such as carbohydrates, lipids, proteins, and nucleic acids are organic compounds that contain carbon atoms. Carbon is considered to be the central atom of organic compounds. Carbon atoms each have four valence electrons and require four more electrons to have a stable outer shell. Due to the repulsion between the valence electrons, the bond sites are all equidistant from each other. This enables carbon to form longs chains and rings. Carbon atoms can form four single covalent bonds with other atoms. For example, methane (CH_4) consists of one carbon atom singly bonded to four separate hydrogen atoms. Carbon atoms can also form double or triple covalent bonds. For example, an oxygen atom can form a double bond with a carbon atom, and a nitrogen atom can form a triple bond with a carbon atom.

Organic and inorganic molecules

Organic molecules always contain carbon. Organic compounds are molecules that are found in or produced by living organisms. Most inorganic molecules do not contain carbon, but some do. Most organic molecules have carbon-hydrogen bonds. Because carbon can form four covalent bonds, organic molecules can be very complex structures. Organic molecules can have carbon backbones that form long chains, branched chains, or even rings. Organic compounds tend to be less soluble in water than inorganic compounds. Organic compounds include four classes: carbohydrates, lipids, proteins, and nucleic acids. Specific examples of organic compounds include sucrose, cholesterol, insulin, and DNA. Inorganic compounds include salts and metals. Specific examples of inorganic molecules include sodium chloride, oxygen, and carbon dioxide.

Adenosine triphosphate (ATP)

Adenosine triphosphate (ATP) is the energy source for most cellular functions. Each ATP molecule is a nucleotide consisting of a central ribose sugar flanked by a purine base and a chain of three phosphate groups. The purine base is adenine, and when adenine is joined to ribose, an adenosine is formed, explaining the name adenosine triphosphate. If one phosphate is removed from the end of the molecule, adenosine diphosphate (ADP) is formed. The molecular formula for ATP is $C_{10}H_{16}N_5O_{13}P_3$, and the condensed structural formula is $C_{10}H_8N_4O_2NH_2(OH)_2(PO_3H)_3H$.

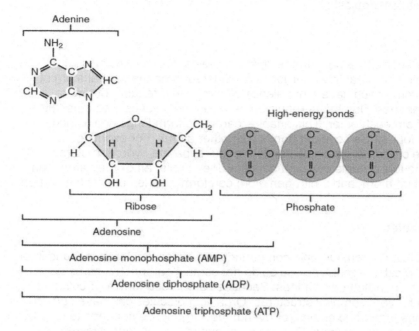

Chemical bonds

Chemical bonds are the attractive forces that bind atoms together forming molecules, and intermolecular forces are the attractive forces between molecules. Chemical bonds include covalent bonds, ionic bonds, and metallic bonds. Covalent bonds are formed from the sharing of electron pairs between two atoms in a molecule. In organic molecules, carbon atoms form single, double, or triple covalent bonds. Organic compounds including proteins, carbohydrates, lipids, and nucleic acids are molecular compounds formed by covalent bonds. Intermolecular forces include hydrogen bonds, London or dispersion forces, and dipole-dipole forces. Hydrogen bonds are the attractive forces between molecules containing hydrogen atoms covalently bonded to oxygen, fluorine, or nitrogen. Hydrogen bonds bind the two strands of a DNA molecule to each other. Two hydrogen bonds join each adenosine and thymine, and three hydrogen bonds join each cytosine and guanine.

Properties of water

Water exhibits numerous properties. Water has a high surface tension due to the cohesion between water molecules from the hydrogen bonds between the molecules. The capillary action of water is also due to this cohesion, and the adhesion of water is due to its polarity. Water is an excellent solvent due to its polarity and is considered the universal solvent. Water exists naturally as a solid, liquid, and gas. The density of water is unusual as it moves between the liquid and solid phases. The density of water decreases as ice freezes and forms crystals in the solid phase. Water is most dense at 4°C. Water can act as an acid or base in chemical reactions. Pure water is an insulator because it has no ions. Water has a high specific heat capacity due to its low molecular mass and bent molecular shape.

Macromolecules

Macromolecules are large molecules made up of smaller organic molecules. Four classes of macromolecules include carbohydrates, nucleic acids, proteins, and lipids. Carbohydrates, proteins, and nucleic acids are polymers or long complex chains of monomers or individual subunits. These polymers are formed when the monomers are joined together in a dehydration process. In this dehydration process, the monomers are joined by a covalent bond and a water molecule is released. Lipids typically are classified as fats, phospholipids, or steroids. The monomers in carbohydrates are simple sugars such as glucose. Polysaccharides are polymers of carbohydrates. The monomers in proteins are amino acids. The amino acids form polypeptide chains, which are folded into proteins. The monomers in nucleic acids are nucleotides. Nucleotides are made up of a sugar, phosphate, and base. Lipids are not actually considered to be polymers. A triglyceride molecule consists of three fatty acids and a glycerol molecule.

Concentration gradients

Concentration gradients, also called diffusion gradients, are differences in the concentration or the number of molecules of solutes in a solution between two regions. A gradient can also result from an unequal distribution of ions across a cell membrane. Solutes move along a concentration gradient by random motion from the region of high concentration toward the region of low concentration in a process called diffusion. Diffusion is the movement of molecules or ions down a concentration gradient. Diffusion is the method by which oxygen, carbon dioxide, and other nonpolar molecules cross a cell membrane. The steepness of the concentration gradient affects the rate of diffusion. Passive transport makes use of concentration gradients as well as electric gradients to move substances across the cell membrane. Active transport can move a substance against its concentration gradient.

Thermodynamics

The first law of thermodynamics states that energy can neither be created nor destroyed. Energy may change forms, but the energy in a closed system is constant. The second law of thermodynamics states that systems tend toward a state of lower energy and greater disorder. This disorder is called entropy. According to the second law of thermodynamics, entropy is increasing. Gibbs free energy is the energy a system that is available or "free" to be released to perform work at a constant temperature. Organisms must be able to use energy to survive. Biological processes such as the chemical reactions involved in metabolism are governed by these laws.

Anabolic and catabolic reactions

Anabolism and catabolism are metabolic processes. Anabolism is essentially the synthesis of large molecules from monomers, whereas catabolism is the decomposition of large molecules into their component monomers. Anabolism uses energy, whereas catabolism produces energy. Anabolism typically builds and repair tissues, and catabolism typically burns stored food to produce energy. For example, protein synthesis, which is the polymerization of amino acids to form proteins, is an anabolic reaction. Mineralization of bones is also an anabolic process. For example, hydrolysis, which is the decomposition of polymers into monomers that releases a water molecule and energy, is a catabolic reaction. Cellular respiration is a catabolic process in which typically glucose combines with oxygen to release energy in the form of adenosine triphosphate (ATP).

Reduction-oxidation reactions

Reduction-oxidation reactions, or redox reactions, involve the transfer of electrons from one substance to another. Reduction occurs in the substance that gains the electrons. Oxidation occurs in the substance that loses the electrons. Cellular respiration and photosynthesis are redox reactions. Cells use the energy stored in food during the redox reaction of cellular respiration. During cellular respiration, glucose molecules are oxidized and oxygen molecules are reduced. Because electrons lose energy when being transferred to oxygen, the electrons are usually first transferred to the coenzyme NAD^+, which is reduced to NADH. The NADH then releases the energy in steps to the oxygen. During photosynthesis, water molecules are split and oxidized and carbon dioxide molecules are reduced. During photosynthesis, when the water molecules are split, electrons are transferred with the hydrogen ions to the carbon dioxide molecules.

Active site structure and substrate binding

Each enzyme has a complex three-dimensional shape that is specifically designed to fit to a particular reactant, which is called the substrate. The enzyme and the substrate join temporarily forming the enzyme-substrate complex. This complex is unstable, and the chemical bonds are likely to be altered to produce a new molecule or molecules. Each enzyme can only combine with specific substrates because of this "lock-and-key" fit. Each enzyme has a designated binding site on the surface that binds to the substrate. Often, the binding site and the active site are at the same location. The enzyme and the substrate are specifically designed for each other, and they are both flexible and can bend and fold to fit into each other as they come together. This concept is referred to as the *induced fit hypothesis*.

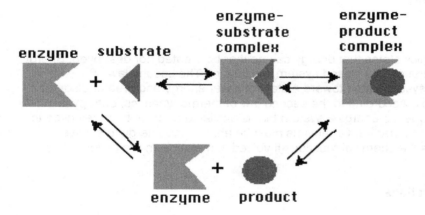

Reaction kinetics

The rate of an enzyme-controlled reaction is affected by many factors in its environment such as temperature, pH, and inhibitors. According to kinetic-molecular theory, increasing the temperature increases the rate of molecular motion. Typically, increasing the temperature increases the rate of these reactions. The optimum temperature is the temperature at which the rate is the fastest and the most product is formed. Increasing the temperature above the optimum temperature actually decreases the reaction rate due to changes on the enzyme's surface that affect its binding to the substrate. The pH also affects enzyme activity due to hydrogen ions binding to the enzyme's surface and changing the enzyme's surface shape. Because enzymes must have a specific shape for their specific substrate, enzymes have a certain pH range in which they can function. Inhibitors are molecules that attach to the enzymes and interfere with the substrates binding to their surfaces, and thus they decrease or even halt enzyme-controlled reactions.

Feedback inhibition

Enzyme-controlled reactions can be regulated by feedback inhibition, or negative feedback. Feedback inhibition can be illustrated by a furnace and thermostat. The inhibitor is this system is the heat. When the furnace runs, the temperature increases. When the temperature reaches a specific level, the thermostat switches the furnace off. When the temperature decreases below a specific level, the thermostat switches the furnace back on, and the cycle begins again. In enzyme-controlled reactions, the end products of a metabolic pathway bind to enzymes found at the beginning of the pathway that generates that product. This causes the reaction rate to decrease. The more product there is, the less product is produced. The less product there is, the more product is produced. This process of feedback inhibition enables a stable range of concentrations that are necessary for homeostasis.

Biochemical pathways

Organisms have different biochemical pathways. Autotrophs that use light to produce energy use photosynthesis as a biochemical pathway. Photosynthesis takes place in chloroplasts in eukaryotic autotrophs. Prokaryotic autotrophs that use inorganic chemical reactions to produce energy use chemosynthesis as a biochemical pathway. Heterotrophs require food and use cellular respiration to release energy from chemical bonds in the molecules of that food. All organisms use cellular respiration to release energy from stored food. Cellular respiration can be aerobic or anaerobic. Most eukaryotes use cellular respiration that takes place in the mitochondria.

Photosynthesis is a food-making process that occurs in three processes: light-capturing events, light-dependent reactions, and light-independent reactions. In light-capturing events, the thylakoids of the chloroplasts, which contain chlorophyll and accessory pigments, absorb light energy and produce excited electrons. Thylakoids also contain enzymes and electron-transport molecules. Molecules involved in this process are arranged in groups called photosystems. In light-dependent reactions, the excited electrons from the light-capturing events are moved by electron transport in a series of steps in which they are used to split water into hydrogen and oxygen ions. The oxygen is released, and the $NADP^+$ bonds with the hydrogen atoms and forms NADPH. ATP is produced from the excited elections. The light-independent reactions use this ATP, NADPH, and carbon dioxide to produce sugars. Three types of photosynthesis are C3, C4, and crassulacean acid metabolism (CAM). In C3 photosynthesis, a three-carbon compound stores the carbon dioxide. In C4 photosynthesis, a four-carbon compound stores the carbon dioxide. In CAM photosynthesis, an acid stores the carbon dioxide. More than 95% of plants perform C3 photosynthesis. C4 photosynthesis can be used by plants in high light regions because it helps conserve water. CAM photosynthesis allows plants to survive long dry spells.

Aerobic cellular respiration is a series of enzyme-controlled chemical reactions in which oxygen reacts with glucose to produce carbon dioxide and water, releasing energy in the form of adenosine triphosphate (ATP). Cellular respiration occurs in a series of three processes: glycolysis, the Krebs cycle, and the electron-transport system. Glycolysis is a series of enzyme-controlled chemical reactions that occur in the cell's cytoplasm. Each glucose molecule is split in half, which releases electrons that are collected by NAD^+ molecules and produces two pyruvic acid molecules and four ATP molecules. Because two ATP molecules are used to split the glucose molecule, glycolysis nets two ATP molecules. The Krebs cycle, also known as the citric acid cycle, is a series of enzyme-controlled chemical reactions that occur in the cell's mitochondria. The Krebs cycle breaks down the pyruvic acid from the glycolysis and releases carbon dioxide and ATP. It also releases electrons, which are collected by NAD^+ and other molecules. These electrons are delivered to the electron-transport system (ETS). The ETS is a series of enzyme-controlled chemical reactions that occurs in the cell's mitochondria. Through a series of reactions in which oxygen atoms are reduced by accepting the electrons from the Krebs cycle, a large amount of ATP is produced. The oxygen ions join with hydrogen ions to produce water. Most of the ATP produced during cellular respiration occurs in the ETS.

Chemosynthesis

Chemosynthesis is the food-making process of chemoautotrophs such as deep-sea-vent microorganisms. Chemosynthesis is unlike photosynthesis in that chemosynthesis does not require light. Chemosynthesis is used by different types of bacteria in extreme environments. Sulfur bacteria live near or in deep-sea vents. Some actually live in other organisms such as huge tube worms near the vents. Hydrogen sulfide is released from deep-sea vents. Instead of sunlight, chemosynthesis uses the energy stored in the chemical bonds of chemicals such as hydrogen sulfide. Carbon is obtained from molecules such as carbon dioxide. During chemosynthesis, the electrons that are removed from the inorganic molecules are combined with carbon possibly from the dissolved carbon dioxide in the seawater or from methane from deep-sea vents to form organic molecules in the form of carbohydrates. Some bacteria use metal ions such as iron and magnesium to obtain the needed electrons. Methanobacteria such as those found in human intestines combine carbon dioxide and hydrogen gas and release methane as a waste product. Nitrogen bacteria such as nitrogen-fixing bacteria in the nodules of legumes convert atmospheric nitrogen into nitrates. In general, chemosynthesis involves the oxidation of inorganic substances.

Prokaryotes and eukaryotes

Cells of the domains of Bacteria and Archaea are prokaryotes. Bacteria cells and Archaea cells are much smaller than cells of eukaryotes. Prokaryote cells are usually only 1 to 2 micrometers in diameter, but eukaryotic cells are usually at least 10 times and possibly 100 times larger than prokaryotic cells. Eukaryotic cells are usually 10 to 100 micrometers in diameter. Most prokaryotes are unicellular organisms, although some prokaryotes live in colonies. Because of their large surface-area-to-volume ratios, prokaryotes have a very high metabolic rate. Eukaryotic cells are much larger than prokaryotic cells. Due to their larger sizes, they have a much smaller surface-area-to-volume ratio and consequently have much lower metabolic rates.

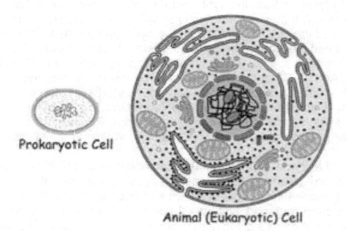

Prokaryotic Cell

Animal (Eukaryotic) Cell

Prokaryotic cells are much simpler than eukaryotic cells. Prokaryote cells do not have a nucleus due to their small size. Their DNA is located in the center of the cell in a region referred to as a nucleoid. Eukaryote cells have a nucleus bound by a double membrane. Eukaryotic cells typically have hundreds or thousands of additional membrane-bound organelles that are independent of the cell membrane. Prokaryotic cells do not have any membrane-bound organelles that are independent of the cell membrane. Once again, this is probably due to the much larger size of the eukaryotic cells. The organelles of eukaryotes give them much higher levels of intracellular division than is possible in prokaryotic cells.

Not all cells have cell walls. Most prokaryotes have cell walls. The cell walls of organisms from the domain Bacteria differ from the cell walls of the organisms from the domain Archaea. Some eukaryotes, such as some fungi, some algae, and plants, have cell walls that differ from the cell walls of the Bacteria and Archaea domains. Most bacteria have cell walls outside of the plasma membrane that contains the molecule peptidoglycan. Peptidoglycan is a large polymer of amino acids and sugars. The peptidoglycan helps maintain the strength of the cell wall. Some of the Archaea cells have cell walls containing the molecule pseudopeptidoglycan, which differs in chemical structure from the peptidoglycan but basically provides the same strength to the cell wall. Some fungi cell walls contain chitin. The cell walls of diatoms, a type of yellow algae, contain silica. Plant cell walls contain cellulose, and woody plants are further strengthened by lignin. Some algae also contain lignin. Animal cells do not have cell walls.

Prokaryote cells have DNA arranged in a circular structure that should not be referred to as a chromosome. Due to the small size of a prokaryote cell, the DNA material is simply located near the center of the cell in a region called the nucleoid. Prokaryote cells lack histone proteins, and therefore the DNA is not actually packaged into chromosomes. Therefore, the DNA molecule located in a prokaryote cell floats freely within the cell. The DNA in a eukaryotic cell is located in the membrane-bound nucleus. Eukaryote cells have linear chromosomes and histone proteins. During mitosis, the chromatin is tightly wound on the histone proteins and packaged as a chromosome. A prokaryotic cell contains only one large DNA molecule, but it also contains tiny rings of DNA called plasmids. Eukaryotic cells may contain several large DNA molecules or chromosomes. Eukaryotes reproduce by mitosis, and prokaryotes reproduce by binary fission.

Selective permeability

The cell membrane, or plasma membrane, has selective permeability with regard to size, charge, and solubility. With regard to molecule size, the cell membrane allows only small molecules to diffuse through it. Oxygen and water molecules are small and typically can pass through the cell membrane. The charge of the ions on the cell's surface also either attracts or repels ions. Ions with like charges are repelled, and ions with opposite charges are attracted to the cell's surface. Molecules that are soluble in phospholipids can usually pass through the cell membrane. Many molecules are not able to diffuse the cell membrane, and, if needed, those molecules must be moved through by active transport and vesicles.

Active and passive transport

Cells can move materials in and out through the cell membrane by active and passive transport. In passive transport, the molecules diffuse across the cell membrane by osmosis. These molecules are moving from a region where they have a high concentration to a region where the concentration is lower. In passive transport, the molecules move across the cell membrane without the cell expending any extra energy. Diffusion and facilitated diffusion are considered passive transport. Facilitated diffusion occurs when molecules are helped across the membrane by certain proteins called channel proteins or carrier proteins. Because facilitated diffusion is still from a region of high to low concentration, it does not require additional energy and is therefore a type of passive transport. In active transport, molecules are forcibly moved from regions where the concentration is low into a region where the concentration is higher. Carrier proteins must carry these ions and molecules, and this requires an expenditure of energy. Some ions are actively pumped across the cell membrane by proteins. Sodium ions are pumped out of cell, and potassium ions are pumped into the cell in this manner.

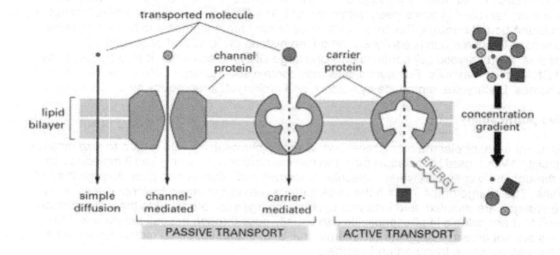

Water movement

Cells must maintain their water balance for homeostasis. If cells have too little water, wastes and poisons can build up in the cells. If cells have too much water, the chemicals in the cells may be diluted. Water is moved in and out of cells by osmosis. Because osmosis is a type of passive transport, the cell cannot actually control this diffusion of water in and out of the cells. The amount of water that diffuses into or out of cells depends on the cell's environment. When the cell's concentration of water and dissolved solids equals that of its environment, the cells are isotonic with their environment. Cells with a lower concentration of water than their environment tend to rapidly gain water by osmosis. These cells are hypotonic with their environment. Cells with a higher concentration of water than their environment tend to rapidly lose water by osmosis. The cells are hypertonic with their environment. If cells are hypotonic or hypertonic, they must expend energy to maintain the proper water balance.

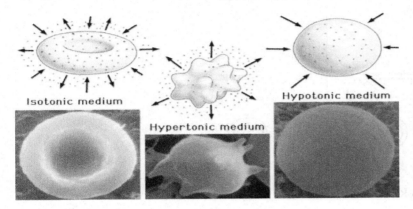

Cell surface proteins and cell communication

In order to maintain a stable internal environment, cells need to send and receive signals from the external environment. Cells have specialized surface proteins called receptors embedded in the cell membrane that allow them to communicate with this external environment. Some surface proteins are exposed to the external side of the membrane. Some surface proteins allow entry to specific needed materials, and others trigger chemical signals inside the cell. Because these proteins have attached carbohydrates, they are called glycoproteins. Due to the cholesterol in the cell membrane, fat-soluble materials can pass straight through the membrane, but water-soluble materials cannot diffuse. Sodium, calcium, and potassium must use these specialized surface proteins to gain entry to the cell. These surface proteins bind to specific chemicals in the materials seeking access to the cell. This triggers a chemical signal to the interior of the cell.

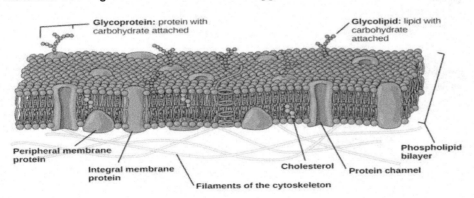

Exocytosis and endocytosis

Larger particles or groups of particles can be transported whole across the cell membrane by being packaged in a piece of cell membrane. Endocytosis is the process by which large particles are moved into the cell, and exocytosis is the process by which large molecules are moves out of the cell. Three main types of endocytosis are phagocytosis, pinocytosis, and receptor-mediated endocytosis. Phagocytosis, or "cell eating," is the process by which large solid particles are engulfed. Pinocytosis, or "cell drinking," is the process by which liquids and dissolved substances are surrounded by small sacs of cell membrane. Receptor-mediated endocytosis is the process by which molecules enter cells through receptor molecules on the cell membrane.

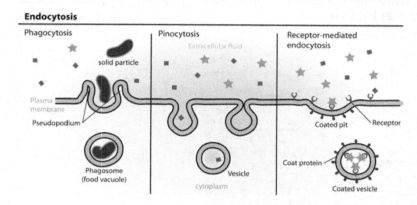

Hormone action and feedback

In order to maintain homeostasis, the endocrine system often employs negative-feedback inhibition or positive-feedback regulation. In negative-feedback inhibition, an increase in an output of a reaction to a stimulus triggers a decrease in the stimulus, which in turn causes a decrease in the original output. In positive-feedback regulation, an increase in an output leads to further increase of the stimulus. An example of negative-feedback inhibition is the release of the hormones insulin and glucagon to maintain the level of glucose in the blood.

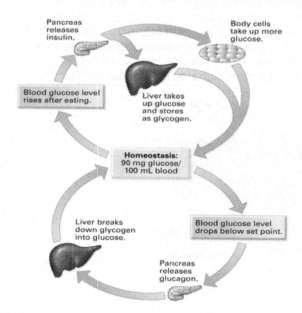

34

Selective permeability

The cell membrane, or plasma membrane, has selective permeability with regard to size, charge, and solubility. With regard to molecule size, the cell membrane allows only small molecules to diffuse through it. Oxygen and water molecules are small and typically can pass through the cell membrane. The charge of the ions on the cell's surface also either attracts or repels ions. Ions with like charges are repelled, and ions with opposite charges are attracted to the cell's surface. Molecules that are soluble in phospholipids can usually pass through the cell membrane. Many molecules are not able to diffuse the cell membrane, and, if needed, those molecules must be moved through by active transport and vesicles.

Active and passive transport

Cells can move materials in and out through the cell membrane by active and passive transport. In passive transport, the molecules diffuse across the cell membrane by osmosis. These molecules are moving from a region where they have a high concentration to a region where the concentration is lower. In passive transport, the molecules move across the cell membrane without the cell expending any extra energy. Diffusion and facilitated diffusion are considered passive transport. Facilitated diffusion occurs when molecules are helped across the membrane by certain proteins called channel proteins or carrier proteins. Because facilitated diffusion is still from a region of high to low concentration, it does not require additional energy and is therefore a type of passive transport. In active transport, molecules are forcibly moved from regions where the concentration is low into a region where the concentration is higher. Carrier proteins must carry these ions and molecules, and this requires an expenditure of energy. Some ions are actively pumped across the cell membrane by proteins. Sodium ions are pumped out of cell, and potassium ions are pumped into the cell in this manner.

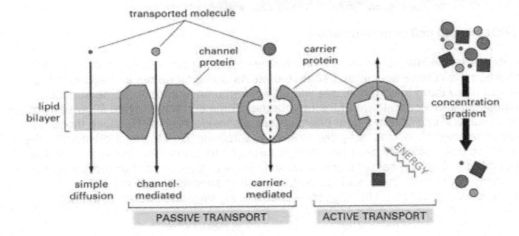

35

Water movement

Cells must maintain their water balance for homeostasis. If cells have too little water, wastes and poisons can build up in the cells. If cells have too much water, the chemicals in the cells may be diluted. Water is moved in and out of cells by osmosis. Because osmosis is a type of passive transport, the cell cannot actually control this diffusion of water in and out of the cells. The amount of water that diffuses into or out of cells depends on the cell's environment. When the cell's concentration of water and dissolved solids equals that of its environment, the cells are isotonic with their environment. Cells with a lower concentration of water than their environment tend to rapidly gain water by osmosis. These cells are hypotonic with their environment. Cells with a higher concentration of water than their environment tend to rapidly lose water by osmosis. The cells are hypertonic with their environment. If cells are hypotonic or hypertonic, they must expend energy to maintain the proper water balance.

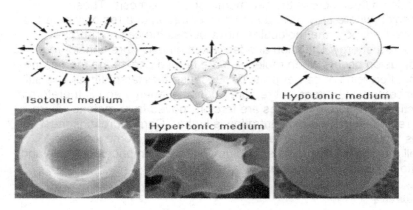

Cell surface proteins and cell communication

In order to maintain a stable internal environment, cells need to send and receive signals from the external environment. Cells have specialized surface proteins called receptors embedded in the cell membrane that allow them to communicate with this external environment. Some surface proteins are exposed to the external side of the membrane. Some surface proteins allow entry to specific needed materials, and others trigger chemical signals inside the cell. Because these proteins have attached carbohydrates, they are called glycoproteins. Due to the cholesterol in the cell membrane, fat-soluble materials can pass straight through the membrane, but water-soluble materials cannot diffuse. Sodium, calcium, and potassium must use these specialized surface proteins to gain entry to the cell. These surface proteins bind to specific chemicals in the materials seeking access to the cell. This triggers a chemical signal to the interior of the cell.

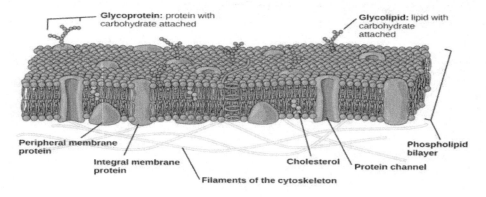

Exocytosis and endocytosis

Larger particles or groups of particles can be transported whole across the cell membrane by being packaged in a piece of cell membrane. Endocytosis is the process by which large particles are moved into the cell, and exocytosis is the process by which large molecules are moves out of the cell. Three main types of endocytosis are phagocytosis, pinocytosis, and receptor-mediated endocytosis. Phagocytosis, or "cell eating," is the process by which large solid particles are engulfed. Pinocytosis, or "cell drinking," is the process by which liquids and dissolved substances are surrounded by small sacs of cell membrane. Receptor-mediated endocytosis is the process by which molecules enter cells through receptor molecules on the cell membrane.

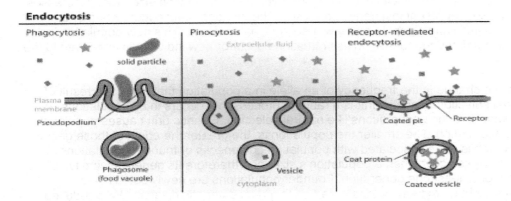

Hormone action and feedback

In order to maintain homeostasis, the endocrine system often employs negative-feedback inhibition or positive-feedback regulation. In negative-feedback inhibition, an increase in an output of a reaction to a stimulus triggers a decrease in the stimulus, which in turn causes a decrease in the original output. In positive-feedback regulation, an increase in an output leads to further increase of the stimulus. An example of negative-feedback inhibition is the release of the hormones insulin and glucagon to maintain the level of glucose in the blood.

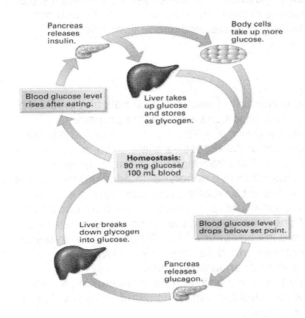

Evolution

Genetic drift

Genetic drift is a microevolutionary process that causes random changes in allele frequencies that are not the result of natural selection. Genetic drift can result in a loss of genetic diversity. Genetic drift greatly impacts small populations. Two special forms of genetic drift are the genetic bottleneck and the founder effect. A genetic bottleneck occurs when there is a drastic reduction in population due to some change such as overhunting, disease, or habitat loss. When a population is greatly reduced in size, many alleles can be lost. Even if the population size greatly increases again, the lost alleles represent lost genetic diversity. The founder effect occurs when one individual or a few individuals populate a new area such as an island. This new population is limited to the alleles of the founder(s) unless mutations occur or new individuals immigrate to the region.

Genetic drift is the change in the frequency of an allele in a population that is not the result of natural or sexual selection. Genetic drift is a random process that occurs in all populations. Genetic drift cannot produce adaptations like natural selection. Genetic drift causes populations to lose genetic information. The smaller the population is, the greater the effect of those genetic losses. Genetic drift is often associated with population bottlenecks or founder populations. Population bottlenecks occur when a population's size, and therefore its genetic diversity, is greatly reduced for at least one generation. Founder populations are newly established populations that might be extremely small, leading to reduced genetic diversity. With reduced genetic variation, a population may not be able to adapt if necessary. Genetic drift may have detrimental effects on endangered species and can lead to extinction.

Gene flow

Gene flow is a microevolutionary process in which alleles enter a population by immigration and leave a population by emigration. Gene flow helps counter genetic drift. When individuals from one genetically distinct population immigrate to a different genetically distinct population, alleles and their genetic information are added to the new population. The added alleles will change the gene frequencies within the population. This increases genetic diversity. If individuals with rare alleles emigrate from a population, the genetic diversity is decreased. Gene flow reduces the genetic differences between populations.

Hardy–Weinberg (HW) equilibrium concept

Hardy–Weinberg (HW) equilibrium is a theoretical concept that uses a mathematical relationship to study gene frequencies. According to HW, if specific conditions are met, the proportions of genotypes in a population can be described by the equation: $p^2 + 2pq + q^2 = 1$, in which p is the frequency of the dominant allele and q is the frequency of the recessive allele. Also, p^2 is the frequency of the homozygous dominant genotype, $2pq$ is the frequency of the heterozygous genotype, and q^2 is the frequency of the homozygous recessive genotype. In addition, the sum of p and q must be equal to one. If the frequencies on the left side of the equation have a sum of one, then the population is in equilibrium, and evolution is not taking place. If the frequencies on the left side of the equation do not have a sum of one, then evolution is taking place. Therefore, the HW equation is only true for populations that are in equilibrium. The HW equilibrium requires the following five conditions to be met: (1) The population must be very large. (2) Mating is random. (3) There are no mutations. (4) No immigration or emigration can occur. (5) All individuals of the population have an equal chance to survive and reproduce. According to this concept, if all five conditions are met, the gene frequencies will remain constant. In reality, these five conditions are rarely met except in a laboratory situation.

Calculating allele frequencies of a simple genetic locus at which there are two alleles (*A* and *a*) in a population of 1,000 individuals given that the population consists of 120 individuals homozygous for the dominant allele (*AA*), 480 heterozygous individuals (*Aa*), and 400 individuals homozygous for the recessive allele (*aa*).

To calculate the frequency of an allele, divide the total number of those alleles in the population by the total number of alleles in the population for that locus as shown in the following equation:

$$\text{allele frequency} = \frac{\text{total \# of allelles in population}}{\text{total \# of alleles in the population for that locus}}.$$

First, find the total number of each type of allele. The 120 *AA* individuals produce 240 *A* alleles. The 480 heterozygous individuals produce 480 *A* alleles and 480 *a* alleles. The 400 *aa* individuals produce 800 *a* alleles. Therefore, there is a total of 720 *A* alleles and 1280 *a* alleles. Adding the 720 and 1,280 yields a total of 2,000 alleles in the population for that locus. The allele frequency for *A* = $\frac{720}{2,000}$ or 0.36. The allele frequency for *a* = $\frac{1,280}{2,000}$ or 0.64.

Natural and artificial selection

Natural selection and artificial selection are both mechanisms of evolution. Natural selection is a process of nature. Natural selection is the way in which a population can change over generations. Every population has variations in individual heritable traits. Not all individuals of a population reproduce. The organisms best suited for survival typically reproduce and pass on their genetic traits. Typically, the more advantageous a trait is, the more common that trait becomes in a population. Natural selection brings about evolutionary adaptations and is responsible for biological diversity. Artificial selection is another mechanism of evolution. Artificial selection is a process brought about by humans. Artificial selection is the selective breeding of domesticated animals and plants such as when farmers choose animals or plants with desirable traits to reproduce. Artificial selection has led to the evolution of farm stock and crops. For example, cauliflower, broccoli, and cabbage all evolved due to artificial selection of the wild mustard plant.

Sexual selection

Sexual selection is a special case of natural selection in animal populations. Sexual selection occurs because some animals are more likely to find mates than other animals. The two main contributors to sexual selection are competition of males and mate selection by females. An example of male competition is in the mating practices of the redwing blackbird. Some males have huge territories and numerous mates that they defend. Other males have small territories, and some even have no mates. An example of mate selection by females is the mating practices of peacocks. Male peacocks display large, colorful tail feathers to attract females. Females are more likely to choose males with the larger, more colorful displays.

Coevolution

Coevolution describes a rare phenomenon in which two populations with a close ecological relationship undergo reciprocal adaptations simultaneously and evolve together, affecting each other's evolution. General examples of coevolution include predator and prey, or plant and pollinator, and parasites and their hosts. A specific example of coevolution is the yucca moths and the yucca plants. Yucca plants can only be pollinated by the yucca moths. The yucca moths lay their eggs in the yucca flowers, and their larvae grow inside the ovary.

Adaptive radiation

Adaptive radiation is an evolutionary process in which a species branches out and adapts and fills numerous unoccupied ecological niches. The adaptations occur relatively quickly, driven by natural selection and resulting in new phenotypes and possibly new species eventually. An example of adaptive radiation is the finches that Darwin studied on the Galápagos Islands. Darwin recorded 13 different varieties of finches, which differed in the size and shape of their beaks. Through the process of natural selection, each type of finch adapted to the specific environment and specifically the food sources of the island to which it belonged. On newly formed islands with many unoccupied ecological niches, the adaptive radiation process occurred quickly due to the lack of competing species and predators.

Molecular evidence that supports evolution

Because all organisms are made up of cells, all organisms are alike on a fundamental level. Cells share similar components, which are made up of molecules. Specifically, all cells contain DNA and RNA. This should indicate that all species descended from a common ancestor. Humans and chimpanzees share approximately 98% of their genes in common, and humans and bacteria share approximately 7% of their genes in common. Humans and zebra fish share approximately 85% of their genes in common. Humans and mustard greens share approximately 15% of their genes in common. Biologists have been able to use DNA sequence comparisons of modern organisms to reconstruct the "root" of the tree of life. Recent discoveries indicate that RNA can both store information and cause itself to be copied, which means that it could produce proteins. Therefore, RNA could have could have evolved first, followed by DNA.

Embryology

Embryos of many organisms pass through several stages, revealing homologies between many species. These embryonic homologies are evidence of a relationship to another species in which the structures emerge past the embryonic stage and are carried into the adult form. For example, in chicken embryos and mammalian embryos, both include a stage in which slits and arches appear in the embryo's neck region that are strikingly similar to gill slits and gill arches in fish embryos. Obviously, adult chickens and adult mammals do not have gills, but this embryonic homology indicates that birds and mammals share a common ancestor with fish. For example, some species of toothless whales have embryos that initially develop teeth that are later absorbed, which indicates that these whales have an ancestor with teeth in the adult form. For example, most tetrapods have five-digit limbs, but birds have three-digit limbs in their wings. However, embryonic birds initially have five-digit limbs in their wings, which develop into a three-digit wing. Tetrapods such as reptiles, mammals, and birds all share a common ancestor with five-digit limbs. In general, embryonic homologies are evidence of a common ancestor.

Homology

Homology is the similarity of structures of different species based on a similar structure in a common evolutionary ancestor. The forelimbs of whales, frogs, horses, lions, humans, bats, and birds all have the same basic pattern of the bones. Specifically, all of these organisms have a humerus, radius, and ulna. Tetrapods all have limbs with five digits at some stage in their development. For example, embryonic birds start with limbs with five digits, but adult bird wings have only three digits. They are all modifications of the same basic evolutionary structure from a common ancestor. Tetrapods resemble the fossils of extinct transitional animal called the *Eusthenopteron*. This would seem to indicate that evolution primarily modifies preexisting structures.

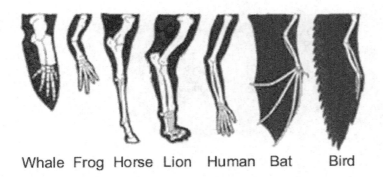

Whale Frog Horse Lion Human Bat Bird

Endosymbiosis theory

The endosymbiosis theory is foundational to evolution. Endosymbiosis provides the path for prokaryotes to give rise to eukaryotes. Specifically, endosymbiosis explains the development of the organelles of mitochondria in animals and chloroplasts in plants. This theory states that some organelles such as mitochondria and chloroplasts in eukaryotic cells originated as free living cells, specifically bacteria. According to this theory, primitive, heterotrophic eukaryotes engulfed smaller, autotrophic bacteria prokaryotes, but the bacteria were not digested. Instead the eukaryotes and the bacteria formed a symbiotic relationship. Eventually, the bacteria transformed into mitochondrion or chloroplasts. Several facts support this theory. Mitochondria and chloroplasts contain their own DNA and can both only arise from other preexisting mitochondria and chloroplasts. The genomes of mitochondria and chloroplasts consist of single, circular DNA molecules with no histones. This is similar to bacteria genomes, not eukaryote genomes. Also, the RNA, ribosomes, and protein synthesis of mitochondria and chloroplasts are remarkably similar to those of bacteria, and both use oxygen to produce ATP. These organelles have a double phospholipid layer that is typical of engulfed bacteria. This theory also involves a secondary endosymbiosis in which the original eukaryotic cells that have engulfed the bacteria are then engulfed themselves by another free-living eukaryote.

Convergent and divergent evolution

Convergent evolution is the evolutionary process in which two or more unrelated species become increasingly similar in appearance. In convergent evolution, natural selection leads to adaptation in these unrelated species belonging to the same kind of environment. For example, the mammals shown below, although found in different parts of the world, developed similar appearances due to their similar environments.

Divergent evolution is the evolutionary process in which organisms of one species become increasingly dissimilar in appearance. As several small adaptations occur due to natural selection, the organisms will finally reach a point at which two new species are formed. Then, these two species will further diverge from each other as they continue to evolve. Adaptive radiation is an example of divergent evolution. Another example is the divergent evolution of the wooly mammoth and the modern elephant from a common ancestor.

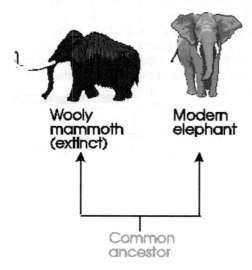

Fossil record

The fossil record provides many types of support for evolution including comparisons from rock layers, transition fossils, and homologies with modern organisms. First, fossils from rock layers from all over the world have been compared, enabling scientists to develop a sequence of life from simple to complex. Based on the fossil record, the geologic timeline chronicles the history of all living things. For example, the fossil record clearly indicates that invertebrates developed before vertebrates and that fish developed before amphibians. Second, numerous transitional fossils have been found. Transitional fossils show an intermediate state between an ancestral form of an organism and the form of its descendants. These fossils show the path of evolutionary change. For example, many transition fossils documenting the evolutionary change from fish to amphibians have been discovered. In 2004, scientists discovered *Tiktaalik roseae*, or the "fishapod," which is a 375-million-year-old fossil that exhibits both fish and amphibian characteristics. For example, scientists have determined that *Pakicetus,* an extinct land mammal, is an early ancestor of modern whales and dolphins based on the specialized structures of the inner ear. Most fossils exhibit homologies with modern organisms. For example, extinct horses are similar to modern horses, indicating a common ancestor.

Cephalization and multicellularity

Two major evolutionary trends are cephalization and multicellularity. Cephalization is the evolutionary trend that can be summarized as "the evolution of the head." In most animals, nerve tissue has been concentrated into a brain at one end of an organism over many generations. Eventually, a head enclosing a brain and housing sensory organs was produced at one end of the organism. Many invertebrates, such as arthropods and annelids and all vertebrates, have undergone cephalization. However, some invertebrates, such as echinoderms and sponges, have not undergone cephalization, and these organisms literally do not have a head. Another evolutionary trend is multicellularity. Life has evolved from simple, single-celled organisms to complex, multicellular organisms. Over millions of years, single-cell organisms gave rise to biofilms, which gave rise to multicellular organisms, which gave rise to all of the major phyla of multicellular organisms present today.

Speciation

Speciation has various modes that bring about the reduced gene flow needed for incipient species to fully separate. Allopatric speciation occurs between two incipient species that become geographically isolated populations. Say, for some geographic reason, two or more groups within a species cannot mate with each other. This could be due to a volcanic eruption, a desert, or a river. Some may still mate, but the gene flow is greatly reduced. Peripatric speciation is a more specific type of allopatric speciation. Peripatric speciation occurs when an extremely small group is geographically isolated at the edge of the rest of the population. This small population size brings about genetic drift relatively quickly. Parapatric speciation occurs within a continuously distributed population. This occurs due to a geographic distance instead of a geographic barrier. Individuals in the population simply choose to mate with close neighbors, leading to a reduced gene flow. Natural selection in the range of those individuals brings about further differentiation among members of the species spread out across the population. Sympatric speciation occurs when a new species develops within a population with no geographic isolation. Sympatric speciation is uncommon but may occur when a species inhabits a new niche.

Biological species are groups of organisms that can breed and produce viable offspring. New species may originate due to reproductive isolation between members of the same species. Prezygotic barriers occur before fertilization and stop or hinder species from mating. Postzygotic barriers occur after fertilization but prevent the hybrids from living or being fertile. If gene flow between two groups of a species is hindered or stopped completely, more significant genetic differences between the two groups can accumulate. In order for speciation to occur, differences between two incipient or emerging species must occur. For the two incipient species to completely split, one of two things must occur. Either mating between the incipient species cannot occur, or the offspring must be nonviable or sterile. These situations can occur in many ways. For example, a change in the mating location or a change in the mating rituals can keep mating from occurring. Changes dues to natural selection or genetic drift may affect mating or viability of offspring. Various modes of geographic and population isolation may occur, but the key to speciation in each type is still the reduced gene flow.

Gradualism

Gradualism is a model of evolutionary rates that states that evolutionary changes occurred slowly or gradually by a divergence of lineages due largely to natural selection. These accumulated changes occurred over millions of years. Many transitional forms occurred between ancestors and modern descendants. Although not all of these transitional forms were preserved in the fossil record, the fossil record clearly supports gradualism. Many transition fossils show adaptations as organisms evolve. The geologic time scale describes this gradual change from simple to complex organisms over millions of years.

Punctuated equilibrium

Punctuated equilibrium is a model of evolutionary rates that states that in some instances, evolutionary changes occurred in relatively short burst that "punctuate" long periods of equilibrium of little or no change. These "short" periods would still consist of hundreds of thousands of years. Most scientists believe that punctuated equilibrium occurred along with gradualism. The fossil record supports punctuated equilibrium for many organisms. Punctuated equilibrium provides an explanation for the supposed numerous "missing links" in the fossil record. If punctuated equilibrium is validated, then there actually are no missing links. Scientists are studying the available data to determine if species do indeed spend most of their lives in an equilibrium state.

Panspermia

The word *panspermia* is a Greek work that means "seeds everywhere." Panspermia is one possible explanation for the origin of life on Earth that states that "seeds" of life exist throughout the universe and can be transferred from one location to another. Three types of panspermia based on the seed-dispersal method have been proposed. Lithopanspermia is described as rocks transferring microorganisms between solar systems. Ballistic panspermia is described as rocks transferring microorganisms within a solar system. Directed panspermia is described as intelligent extraterrestrials spreading the seeds to other planets and solar systems. The panspermia hypothesis only proposes the origin of life on Earth. It does not offer an explanation for the origin of life in the universe or explain the origin of the seeds themselves.

Abiotic synthesis of organic compounds and the Miller–Urey experiment

Scientists have performed sophisticated experiments to determine how the first organic compounds appeared on Earth. First, scientists performed controlled experiments that closely resembled the conditions similar to an early Earth. In the classic Miller–Urey experiment (1953), the Earth's early atmosphere was simulated with water, methane, ammonia, and hydrogen that were stimulated by an electric discharge. The Miller–Urey experiment produced complex organic compounds including several amino acids, sugars, and hydrocarbons. More recently, scientists have been able to more accurately simulate the atmospheric conditions of early Earth and still produced complex organic molecules. Later experiments by other scientists produced nucleic acids. Recently, Jeffrey Bada, a former student of Miller, was able to produce amino acids in a simulation using the Earth's current atmospheric conditions with the addition of iron and carbonate to the simulation. This is significant because in previous studies using Earth's current atmosphere, the amino acids were destroyed by the nitrites produced by the nitrogen.

Biological influence on atmospheric composition

The early atmosphere of Earth had little or possibly no oxygen. Early rocks had high levels of iron at their surfaces. Without oxygen, the iron just entered into the early oceans as ions. In the same time frame, early photosynthetic algae was beginning to grow abundantly in the early ocean. During photosynthesis, the algae would produce oxygen gas, which oxidized the iron at the rocks' surfaces, forming an iron oxide. This process basically kept the algae in an oxygen-free environment. As the algae population grew much larger, it eventually produced such a large amount of oxygen that it could not be removed by the iron in the rocks. Because the algae at this time were intolerant to oxygen, the algae became extinct. Over time, a new iron-rich layer of sediments formed, and algae populations reformed, and the cycle began again. This cycle repeated itself for millions of years. Iron-rich layers of sediment alternated with iron-poor layers. Gradually, algae and other life forms evolved that were tolerant to oxygen, stabilizing the oxygen concentration in the atmosphere at levels similar to those of today.

Self-replication

Several theories for the origin of life involve the self-replication of molecules. In order for life to have originated on Earth, proteins and RNA must have been replicated. Theories propose mechanisms from which bacteria may have evolved from nonliving chemicals. Theories that combine the replication of proteins and RNA seem promising. One such theory is called RNA world. RNA world explains how the pathway of DNA to RNA to protein may have originated by proposing the reverse process. In RNA world, RNA is the precursor to DNA. Scientists have shown that RNA can actually function both as a gene and as an enzyme. Also, RNA can be transcribed into DNA. In RNA world, RNA molecules self-replicated and evolved through recombination and mutations. RNA molecules developed the ability to act as enzymes. Eventually, RNA began to synthesize proteins. Finally, DNA molecules were copied from the RNA in a process of reverse transcription.

Extinction of a species

Genetic diversity provides a mechanism for populations to adapt to changing environments or even human impacts. With a diverse genome, individuals possessing genes making them better suited for the environment are more likely to exist. Without genetic diversity, populations cannot develop adaptations. Populations cannot resist diseases or adapt to changes in the habitat. As the populations of endangered species decrease, genetic diversity decreases even further. Normally, natural selection selects genes that resist diseases or help the organism to adapt to changes in the habitat, but if those genes have drifted out of the population, the population cannot evolve and may become extinct. A small gene pool does not provide much variety for selection. For example, tigers in India are now in danger of extinction. Studies show that more than 90% of the genome has been lost largely due to a period when tigers were heavily killed by British officials and Indian royalty. With fewer than 2,000 tigers in the world and these in small populations, the genetic diversity can continue to decrease, possibly leading to extinction if much effort is not made to preserve the remaining genetic diversity.

A changing environment may lead to the extinction of a species. If an animal has a small tolerance range to food sources and habitat needs or if a population is small, it is less likely to adapt to changes in the environment. Climate change and global warming can affect an ecosystem. Some species may not be able to adapt even to seemingly minor temperature changes especially if their populations are small. Animals needing cooler climates may need to move to cooler habitats. Melting ice caps and glaciers and rising sea levels can seriously disrupt many ecosystems and affect numerous species. For example, the giant panda feeds almost exclusively on bamboo. Bamboo is being threatened by global warming. Due to the dwindling of their food source, giant pandas are less able to adapt to a changing environment. For example, the polar bear may become extinct due to global warming as the polar bears' habitat is being destroyed. Sea turtles may become extinct as the rising sea levels destroy the beaches needed for egg laying. Even if the beaches are not destroyed, increasing temperatures affect the incubation process and the number of offspring being produced.

Humans are responsible for impacting the environment in such a way as to endanger or harm species that may even lead to extinction. Humans destroy habitats directly through deforestation and clearing land for agriculture, logging, mining, and urbanization. Humans also threaten or endanger species through overfishing and overhunting. Pollution can destroy a habitat, and if a species is unable to relocate, this can cause extinction. Introduction of an invasive species that introduces a new predator or competitor to the ecosystem can cause extinction. An example of human impacts leading to the extinction of a species is the case of the passenger pigeon. Millions of passenger pigeons were killed for meat from around 1850 to 1880. Because passenger pigeons only lay one egg at a time, huge flocks were destroyed. The last passenger pigeon died in 1914.

Interspecific competition is competition between individuals of different species for the same limited resources such as food, water, sunlight, and living space. This is especially threatening if the two species share a limiting resource and that resource is not in abundant supply. Interspecific competition can limit the population size of a species. With reduced population side, there is less genetic variation. The species may not be able to adapt to environment of other changes in the ecosystem. For example, firs and spruces compete for resources in coniferous forests. Cheetahs and lions compete for prey in savannas.

DNA and RNA

DNA and RNA are both nucleic acids composed of nucleotides made up of a sugar, a base, and a phosphate molecule. DNA and RNA have three of their four bases in common: guanine, cytosine, and adenine. DNA contains the base thymine, but RNA replaces thymine with uracil. DNA is deoxyribonucleic acid. RNA is ribonucleic acid. DNA is located in the nucleus and mitochondria. RNA is found in the nucleus, ribosomes, and cytoplasm. DNA contains the sugar deoxyribose, and RNA contains the sugar ribose. DNA is double stranded, but RNA is single stranded. DNA has the shape of a double helix, but RNA is complexly folded. DNA contains the genetic blueprint and instructions for the cell. RNA carries out those instructions with its various forms. Messenger RNA, mRNA, is a working copy of DNA, and transfer RNA, tRNA, collects the needed amino acids for the ribosomes during the assembling of proteins. Ribosomal RNA, rRNA, forms the structure of the ribosomes.

Sugar-phosphate backbone

The DNA molecule consists of two strands in the shape of double helix, which resembles a twisted "ladder." The "rungs" of the ladder consist of complementary base pairs of the nucleotides. The "legs" of the ladder consist of chains of nucleotides joined by the bond between the phosphate and sugar molecules. In DNA, the sugar is deoxyribose. The sugars and phosphates are joined together by covalent bonds. The RNA molecule has one strand instead of two. In RNA, the sugar is ribose. The sugars and phosphates of the nucleotides are joined in the same way.

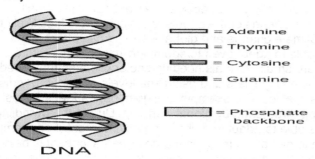

Complementary base pairing

According to Chargaff's rule, DNA always has a 1:1 ratio of purine to pyrimidine. The amount of adenine always equals the amount of thymine, and the amount of guanine always equals the amount of cytosine. DNA contains the bases guanine, cytosine, thymine, and adenine. RNA also contains guanine, cytosine, and adenine, but thymine is replaced with uracil. In DNA, adenine always pairs with thymine, and guanine always pairs with cytosine. In RNA, adenine always pairs with uracil, and guanine always pairs with cytosine. The pairs are bonded together with hydrogen bonds.

Chromosome structure

Prokaryotes contain one DNA molecule. These molecules are usually arranged in a ring that contains all of their genes. Eukaryotes have multiple chromosomes, each containing one DNA molecule. These chromosomes are linear in appearance. Eukaryotic and Archaea bacteria have chromatin that consists of DNA, protein, and RNA. Bacteria do not have histones in their protein. Chromatin is arranged in nucleosomes. A nucleosome is a histone complex with 146 nucleotide pairs wrapped around the histones. The nucleosomes are strung together by a string a DNA. This string of chromosomes coils to form the chromatin. During mitosis, the string is looped and compactly folded to form the chromosome. Eukaryotic chromosomes have telomeres located at their tips. Telomeres are repetitive sequences of DNA that maintain the ends of the linear chromosomes and keep those ends from deteriorating.

DNA replication

DNA replication begins when the double strands of the parent DNA molecule are unwound and unzipped. The enzyme helicase separates the two strands by breaking the hydrogen bonds between the base pairs that make up the rungs of the twisted ladder. These two single strands of DNA are called the replication fork. Each separate DNA strand provides a template for the complementary DNA bases, G with C and A with T. The enzyme DNA polymerase aids in binding the new base pairs together. Short segments of DNA called Okazaki fragments are synthesized with the lagging strand with the aid of RNA primase. At the end of this process, part of the telomere is removed. Then, enzymes check for any errors in the code and make repairs. This results in two daughter DNA molecules each with half of the original DNA molecule that was used as a template.

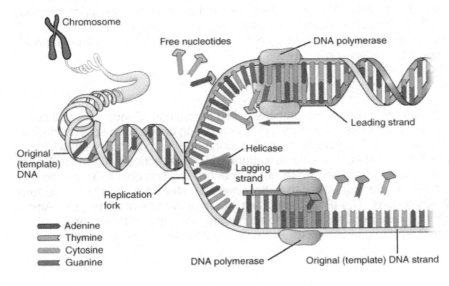

47

RNA transcription

Transcription is the process by which a segment of DNA is copied onto a working blueprint called RNA. Each gene has a special region called a promoter that guides the beginning of the transcription process. RNA polymerase unwinds the DNA at the promoter of the needed gene. After the DNA is unwound, one strand or template is copied by the RNA polymerase by adding the complementary nucleotides, G with C , C with G, T with A, and A with U. Then, the sugar phosphate backbone forms with the aid of RNA polymerase. Finally, the hydrogen bonds joining the strands of DNA and RNA together are broken. This forms a single strand of messenger RNA or mRNA.

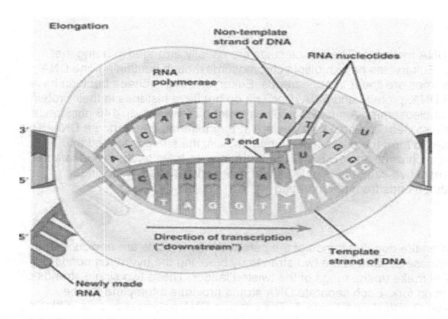

mRNA processing

After mRNA is transcribed, it must be processed. First, during transcription a cap is added. Needed chemicals are readied at the signal sites for cleavage where the polyadenylation will occur. After cleavage, the poly A tail starts to form. The final mRNA strand then with a cap and poly A tail, is ready for splicing. This mRNA strand, which is called the primary transcript, contains one intron and two exons. Once the spliceosome is formed, the intron is cleaved and the two exons are ligated together. After splicing, the working components involved in the splice degrade, and the mRNA strand is ready for translation.

Translation

Ribosomes synthesize proteins from mRNA in a process called translation. Sequences of three amino acids called codons make up the strand of mRNA. Each codon codes for a specific amino acid. The ribosome is composed of two subunits, a larger subunit and a smaller subunit, which are composed of ribosomal RNA (rRNA). The smaller subunit of RNA attaches to the mRNA near the cap. The smaller subunit slides along the mRNA until it reaches the first codon. Then, the larger subunit clamps onto the smaller subunit of the ribosome. Transfer RNA (tRNA) has codons complementary to the mRNA codons. The tRNA molecules attach at the site of translation. Amino acids are joined together by peptide bonds. The ribosome moves along the mRNA strand repeating this process until the protein is complete. Proteins are polymers of amino acids joined by peptide bonds.

Genetics and Information Transfer

Independent assortment

Mendel's law of independent assortment states that alleles of one characteristic separate independently of the alleles of another characteristic. This means that traits are transmitted independently of each other. This can be shown in dihybrid crosses. For example, in pea plants, tall plants (T) are dominant over short plants (t), and axial flowers (A) are dominant over terminal flowers (a). In a genetic cross of two pea plants, one homozygous tall and axial (TTAA), and one homozygous short and terminal (ttaa), the F_1 generation will be 100% heterozygous tall and axial flowers (TtAa). If these F_1 plants are crossed, the resulting F_2 generation is shown below. There are nine genotypes for tall plants with axial flowers: one TTAA, two TTAa, two TtAA, and four TtAa. There are three genotypes for tall plants with terminal flowers: one TTaa and two Ttaa. There are three genotypes for short plants with axial flowers: one ttAA and two ttAa. There is only one genotype for short plants with terminal flowers: ttaa. This cross has a 9:3:3:1 ratio.

	TA	Ta	tA	ta
TA	TTAA	TTAa	TtAA	TtAa
Ta	TTAa	TTaa	TtAa	Ttaa
tA	TtAA	TtAA	ttAA	ttAa
ta	TtAa	Ttaa	ttAa	ttaa

Law of segregation

Mendel's law of segregation is associated with monohybrid crosses. The law of segregation states that the alleles for a trait separate when gametes are formed, which means that only one of the pair of alleles for a given trait is passed to the gamete. This can be shown in monohybrid crosses. For example, in pea plants, tall plants (T) are dominant over short plants (t). In a genetic cross of two pea plants that are homozygous for plant height, the F_1 generation will be 100% heterozygous tall plants.

	t	t
T	Tt	Tt
T	Tt	Tt

If the heterozygous tall plants are crossed, the F_2 generation should be 50% heterozygous tall, 25% homozygous tall, and 25% homozygous short.

	T	t
T	TT	Tt
t	Tt	tt

Monohybrid crosses

A monohybrid cross is a genetic cross for a single trait that has two alleles. A monohybrid cross can be used to show which allele is dominant for a single trait. The first monohybrid cross typically occurs between two homozygous parents. Each parent is homozygous for a separate allele for a particular trait. For example, in pea plants, green pods (G) are dominant over yellow pods (g). In a genetic cross of two pea plants that are homozygous for pod color, the F_1 generation will be 100% heterozygous green pods.

	g	g
G	Gg	Gg
G	Gg	Gg

If the plants with the heterozygous green pods are crossed, the F_2 generation should be 50% heterozygous green, 25% homozygous green, and 25% homozygous yellow.

	G	g
G	GG	Gg
g	Gg	gg

Dihybrid crosses

A dihybrid cross is a genetic cross for two traits that each have two alleles. For example, in pea plants, green pods (G) are dominant over yellow pods (g), and yellow seeds (Y) are dominant over green seeds (y). In a genetic cross of two pea plants that are homozygous for pod color and seed color, the F_1 generation will be 100% heterozygous green pods and yellow seeds (GgYy). If these F_1 plants are crossed, the resulting F_2 generation is shown below. There are nine genotypes for green-pod, yellow-seed plants: one GGYY, two GGYy, two GgYY, and four GgYy. There are three genotypes for green-pod, yellow-seed plants: one GGyy and two Ggyy. There are three genotypes for yellow-pod, yellow-seed plants: one ggYY and two ggYy. There is only one genotype for yellow-pod, yellow-seed plants: ggyy. This cross has a 9:3:3:1 ratio.

	GY	Gy	gY	gy
GY	GGYY	GGYy	GgYY	GgYy
Gy	GGYy	GGyy	GgYy	Ggyy
gY	GgYY	GgYY	ggYY	ggYy
gy	GgYy	Ggyy	ggYy	ggyy

Pedigree

Pedigree analysis is a type of genetic analysis in which an inherited trait is studied and traced through several generations of a family to determine how that trait is inherited. A pedigree is a chart arranged as a type of family tree using symbols for people and lines to represent the relationships between those people. Squares usually represent males, and circles represent females. Horizontal lines represent a male and female mating, and the vertical lines beneath them represent their children. Usually, family members who possess the trait are fully shaded and those that are carriers only of the trait are half-shaded. Genotypes and phenotypes are determined for each individual if possible. The pedigree below shows the family tree of a family in which the first male who was red-green color blind mated with the first female who was unaffected. They had five children. The three sons were unaffected, and the two daughters were carriers.

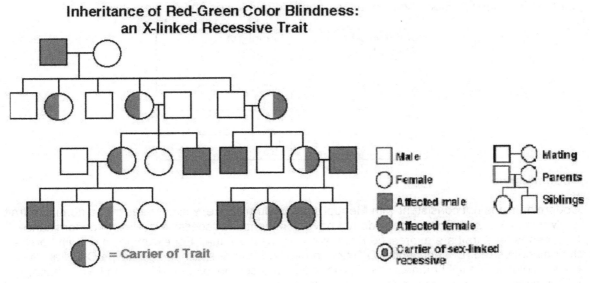

Inheritance of Red-Green Color Blindness: an X-linked Recessive Trait

Linkage

Linkage is an exception to Mendel's law of independent assortment, which states that two genes will assort independently and randomly from each other. Linkage can occur when two genes are located on the same chromosome. Each chromosome has several genes, and those genes tend to be inherited together. Genes that are located on the same chromosome and tend to be inherited together are called linkage groups. Because the genes are on the same chromosome, they do not separate during meiosis. During meiosis, the genes in a linkage group always go into the same gamete together. Due to linkage, genes with different characteristics are inherited together more frequently than is predicted using the laws of probability. An example of a linkage group is found on chromosome number 4. This linkage group includes genes for Parkinson's disease, narcolepsy, and Huntington's disease.

Sex-linked inheritance

Sex-linked inheritance is an exception to Mendel's law of independent assortment, which states that two genes will assort independently and randomly from each. In human genetics, females have two X chromosomes, and males have one X and one Y chromosome. Sex-linked traits are carried on the X chromosome. Because females have two X chromosomes, they have two copies of genes found on the X chromosome. Females may possess a recessive allele for various disorders on one X chromosome, but as long as they possess the dominant allele for normal functioning on the other X chromosome, they will not have the disorder. Females who are heterozygous for a trait such as color blindness or hemophilia are only carriers. Because males have only one X chromosome, if they possess the recessive allele for a disorder, it will be expressed. Examples of traits that are a result of sex linkage are color blindness, hemophilia, a form of muscular dystrophy, and some forms of anemia.

Multiple alleles

Multiple alleles result in a type of non-Mendelian inheritance. In Mendelian inheritance, only two alleles for each gene exist. For example, Mendel's pea plants were either tall or short. Mendel's pea plants had either yellow or green pods. Often, there are more than two possibilities for a particular trait. For example, in human genetics, blood type has many variations. Multiple allele inheritance occurs where there are more than two different alleles of a gene for a particular trait. Even though there may be several alleles for a particular trait, each individual can still only possess two of those possible alleles. For example, the three human blood alleles are I^A (blood contains type A antigens), I^B (blood contains type B antigens), and i (blood contains neither type A nor type B antigens). Blood types and their possible genotypes are shown below.

Blood Type (Phenotype)	Genotype
A	$I^A I^A$, $I^A I^O$
B	$I^B I^B$, $I^B I^O$
AB	$I^A I^B$
O	ii

Codominance

Codominance is not consistent with Mendel's law of dominance, which states that a dominant trait is always expressed. Some genes do not have dominant and recessive alleles. In codominance, both genes for a trait are expressed in a heterozygous individual. For example, in certain horses, the hair colors red (D^R) and white (D^W) are codominant. Horses with the genotype $D^R D^G$ are a golden color with a light mane and tail. Their coats are composed of red hairs and white hairs, which when mixed together causes their coats to have a golden color.

Incomplete dominance

Incomplete dominance is not consistent with Mendel's law of dominance, which states that a dominant trait is always expressed. In these situations, there is no dominant or recessive allele. Instead, both alleles are expressed, and when they are, they blend or mix. For example, there are two alleles for petal color for snapdragons: red (C^R) and white (C^W). Crossing a red snapdragon with a white snapdragon yields an F_1 generation that is 100% heterozygous pink. Crossing two pink snapdragons yields an F_2 generation that is 50% heterozygous pink, 25% homozygous white, and 25% homozygous red.

Polygenic inheritance

Polygenic inheritance occurs when a trait is determined by the interaction of many different genes. In Mendelian genetics, traits are determined by just one pair of genes with two alleles. For example, Mendel's pea plants had either red or white flowers, and his plants were either tall or short. An example of polygenic genetic inheritance in human genetics is skin color and height. Each is controlled by at least four pairs of genes. Skin color has many variations between very light and very dark. Height has many variations between very short and very tall. Eye color and intelligence are also polygenic.

Epistasis

Epistasis occurs when a gene at one locus inhibits the expression of a gene at another locus. For example, in mice, black hair (B) is dominant over brown hair (b). But a different gene at a different locus determines whether or not pigment is deposited (C for deposited and c for not deposited) on the mouse hair. A black mouse with genotypes BB or Bb will only have that color deposited if the pigment genotype is CC or Cc. A black mouse with genotype BBcc will be white. A brown mouse can have genotypes bbCC or bbCc, but a mouse with genotype bbcc will also be white.

Pleiotropy

In Mendelian inheritance, each gene can influence only one trait. Most genes can affect many traits or have multiple phenotypes. Pleiotropy is the situation in which one gene influences several seemingly unrelated traits. Therefore, a gene that affects multiple traits is pleiotropic. Pleiotropy can be due to normal or mutated genes. Genes code for proteins. Because proteins are often used in more than one tissue or more than one area of the body, a missing protein can cause many complications. For example, the hormone insulin is a protein. If the insulin receptors are faulty, then the cells cannot recognize and use the insulin. Other examples of pleiotropy include inherited diseases such as cystic fibrosis, sickle-cell anemia, phenylketonuria (PKU), and albinism.

Mitochondrial inheritance

Mitochondria are cellular organelles that produce energy for the cell. Mitochondria contain their own DNA consisting of 37 genes arranged in a circular structure. Mitochondrial DNA is transmitted maternally, which means that these mitochondrial genes are only inherited from the mother. The offspring's mitochondria only come from the oocyte (egg cell), not from the sperm. Sperm cells only contain mitochondria in their tails, which does not enter the egg during fertilization. Mitochondrial inheritance is not consistent with Mendelian inheritance in which the zygote derives half of the genetic material from the mother and half from the father. Most of these genes code for proteins related to muscular disorders.

Chromosomal aberrations

Chromosomal aberrations are changes in DNA sequences on the chromosomal level. These mutations typically involve many genes and often result in miscarriages. Chromosomal aberrations include translocations, deletions, inversions, and duplications. Translocations occur when a piece of DNA breaks off of one chromosome and is joined to another chromosome. Deletions occur when a piece of DNA breaks off on a chromosome and is lost without reattaching. Inversions occur when a piece of DNA breaks off of one chromosome and becomes reattached to that same chromosome but with an inverted or flipped orientation. Duplications occur when a piece of DNA is replicated and attached to the original piece of DNA in sequence.

Down syndrome

Down syndrome is a type of aneuploidy in which an individual has an abnormal number of chromosomes. Down syndrome, also known as trisomy 21, is caused by a trisomy of the 21st chromosome as shown in the karyotype below. When a gamete with an extra 21st chromosome unites with a normal gamete, the result is a group of three chromosomes instead of a diploid set. A trisomy can occur during nondisjunction. Nondisjunction occurs if a pair of chromosomes fails to separate during meiosis in the formation of an egg or sperm cell.

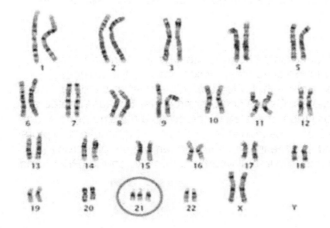

Sickle-cell anemia

Sickle-cell anemia is a genetic disorder that is the result of a gene mutation. Specifically, sickle-cell anemia is the result of the point mutation in which just one nucleotide is changed. The point mutation is a substitution in which adenine is substituted for thymine in the DNA molecule. This results in a defective form of hemoglobin. Sickle-cell anemia occurs when a person is homozygous for the defective gene. Sickle-cell trait occurs when a person is heterozygous for the defective gene, and this person is a carrier but usually suffers no ill effects. Sickle-cell anemia is an example of pleiotropy, in which a change in one gene affects multiple aspects of a person's health. These health problems are due to the abnormally sickle-shaped red blood cells that block the flow of blood, damaging tissues and organs. The sickle-shaped cells tend to rupture, leading to anemia.

Sickle cell anemia

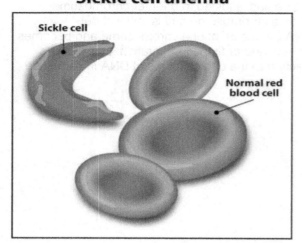

Mutations

Mutations are one of the main sources of genetic variation. Mutations are changes in DNA. The changes can be gene mutations such as the point mutations of substitution, addition, or deletion, or the changes can be on the chromosomal level such as the chromosomal aberrations of translocations, deletions, inversions, and duplications. Mutations are random and can benefit, harm, or have no effect on the individual. Somatic mutations do not affect inheritance and therefore do not affect genetic variation with regard to evolution. Germline mutations that occur in gametes (eggs and sperm) can be passed to offspring and therefore are very important to genetic variation and evolution. Mutations introduce new genetic information into the genome.

Crossing over

Crossing over is a major source of genetic variation. Crossing over is the exchange of equivalent segments of DNA between homologous chromosomes. Crossing over occurs during meiosis in prophase I. During synapsis, a tetrad is formed when homologous chromosomes pair up. Also during synapsis, the chromatids are extremely close together and sometimes the chromatids swap genes. Because genes have more than one allele, this allows for an exchange of genetic information. Crossing over is that exchange of genes. Crossing over can occur several times along the length of the chromosomes. Although crossing over does not introduce new information, it does introduce new combinations of the information that is available. Without crossing over during meiosis, only two genetically different gametes can be formed. With just one instance of crossing over, four genetically different gametes can be formed. With crossing over, each gamete contains genes from both the father and the mother. Crossing over leads to variation in traits among gametes, which leads to variation in traits among offspring.

Genetic exchange

Genetic exchange, or the transfer of DNA from one organism to another, is a source of genetic variation. Three general types of genetic exchange are transduction, transformation, and conjugation. Transduction occurs when genetic material is transferred from one bacterium to another by a bacteriophage. A bacteriophage is a virus that infects a bacterium. As the new bacteriophages are replicated, some of the host bacteria DNA can be added to the virus particles. Transformation occurs when a cell obtains new genetic information from its environment or surroundings. Many bacteria take up DNA fragments such as plasmids from their surroundings to obtain new genes. Conjugation occurs when bacteria or single-celled organisms are in direct contact with each other. Genes can be transferred from one into the other while the two cells are joined.

Independent assortment during sexual reproduction

Independent assortment during sexual reproduction is a source of genetic variation. Mutations originally brought about changes in DNA leading to alleles or different forms of the same gene. During sexual reproduction, these allele are "shuffled" or "independently sorted," producing individuals with unique combinations of traits. Gametes are produced during meiosis, which consists of two cell divisions: meiosis I and meiosis II. Meiosis I is a reduction division in which the diploid parent cell divides into two haploid daughter cells. During the metaphase of meiosis I, the homologous pairs (one from the mother and one from the father) align on the equatorial plane. The orientation of the homologous pairs is random, and each placement is independent of another's placement. The number of possible arrangements increases exponentially as the number of chromosomes increases. The independent assortment of chromosomes during metaphase in meiosis I provides a variety of gametes with tremendous differences in their combinations of chromosomes.

Cell cycle

The cell cycle consists of three stages: interphase, mitosis, and cytokinesis. Interphase is the longest stage of the cell cycle. Cells typically spend more than 90% of the cell cycle in interphase. Interphase includes two growth phases called G1 and G2. The order of interphase is the first growth cycle, GAP 1 (G1 phase), followed by the synthesis phase (S), followed by the second growth phase, GAP 2 (G2 phase). During the G1 phase of interphase, the cell increases the number of organelles by forming diploid cells. During the S phase of interphase, the DNA is replicated, and the chromosomes are doubled. During the G2 phase of interphase, the cell synthesizes needed proteins and continues to increase in size. During mitosis, the cell completes four phases: prophase, metaphase, anaphase, and telophase. During mitosis, the two sets of DNA that are arranged as the duplicated chromosomes are separated. Organelles such as chloroplasts and mitochondria also divide. During cytokinesis, the parent cell divides to form two identical daughter cells. After cytokinesis, the daughter cells begin interphase and the cell cycle starts again.

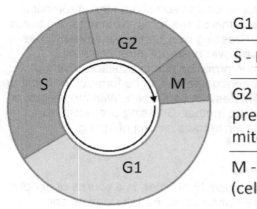

G1 - Growth

S - DNA synthesis

G2 - Growth and preparation for mitosis

M - Mitosis (cell division)

Mitosis

Mitosis is the asexual process of cell division. During mitosis, one parent cell divides into two identical daughter cells. Mitosis is used for growth, repair, and replacement of cells. Some unicellular organisms reproduce asexually by mitosis. Some multicellular organisms can reproduce by fragmentation or budding, which involves mitosis. Mitosis consists of four phases: prophase, metaphase, anaphase, and telophase. During prophase, the spindle fibers appear, and the DNA is condensed and packaged as chromosomes that become visible. The nuclear membrane breaks down, and the nucleolus disappears. During metaphase, the spindle apparatus is formed and the centromeres of the chromosomes line up on the equatorial plane. During anaphase, the centromeres divide and the two chromatids separate and are pulled toward the opposite poles of the cell. During telophase, the spindle fibers disappear, the nuclear membrane reforms, and the DNA in the chromatids is decondensed.

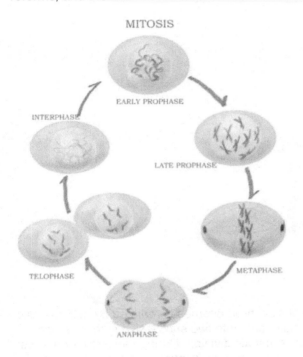

Cytokinesis

Cytokinesis is the dividing of the cytoplasm and cell membrane by the pinching of a cell into two new daughter cells at the end of mitosis. This occurs at the end of telophase when the actin filaments in the cytoskeleton form a contractile ring that narrows and divides the cell. In plant cells, a cell plate forms across the phragmoplast, which is the center of the spindle apparatus. In animal cells, as the contractile ring narrows, the cleavage furrow forms. Eventually, the contractile ring narrows down to the spindle apparatus joining the two cells and the cells eventually divide. Photos of the cell plate of a plant cell by transmission electron microscopy (TEM) and the cleavage furrow of an animal cell by scanning electron microscopy (SEM) are shown below.

(a) Animal cell

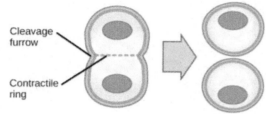

(b) Plant cell

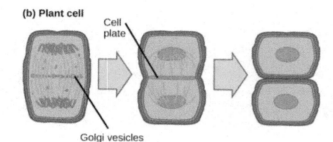

Meiosis

Meiosis is a type of cell division in which the number of chromosomes is reduced by half. Meiosis produces gametes, or egg and sperm cells. Meiosis occurs into two successive stages, which consist of a first mitotic division followed by a second mitotic division. During meiosis I, or the first meiotic division, the cell replicates its DNA in interphase and then continues through prophase I, metaphase I, anaphase I, and telophase I. At the end of meiosis I, there are two daughter cells that have the same number of chromosomes as the parent cell. During meiosis II, the cell enters a brief interphase but does not replicate its DNA. Then, the cell continues through prophase II, metaphase II, anaphase II, and telophase II. During prophase II, the unduplicated chromosomes split. At the end of telophase II, there are four daughter cells that have half the number of chromosomes as the parent cell.

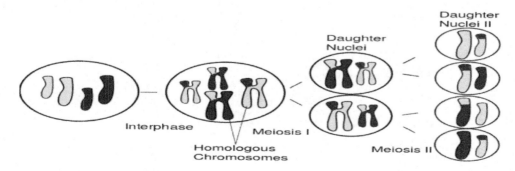

Cell cycle checkpoints

During the cell cycle, the cell goes through three checkpoints to ensure that the cell is dividing properly at each phase, that it is the appropriate time for division, and that the cell has not been damaged. The first checkpoint is at the end of the G1 phase just before the cell undergoes the S phase, or synthesis. At this checkpoint, a cell may continue with cell division, delay the division, or rest. This resting phase is called G0. In animal cells, the G1 checkpoint is called restriction. Proteins called cyclin D and cyclin E, which are dependent on enzymes cyclin-dependent kinase 4 and cyclin-dependent kinase 2 (CDK4 and CDK2), respectively, largely control this first checkpoint. The second checkpoint is at the end of the G2 phase just before the cell begins prophase during mitosis. The protein cyclin A, which is dependent on the enzyme CDK2, largely controls this checkpoint. During mitosis, the third checkpoint occurs at metaphase to check that the chromosomes are lined up along the equatorial plane. This checkpoint is largely controlled by cyclin B, which is dependent upon the enzyme CDK1.

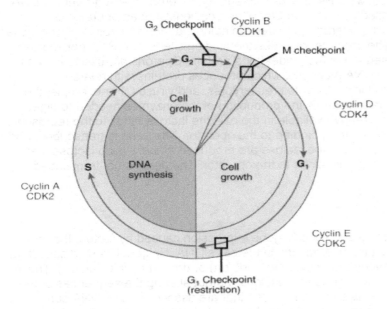

Promoters

Promoters are DNA sequences typically 100 to 1,000 base pairs in length that are usually located upstream of the gene needed for transcription. Basically, promoters signal the beginning of transcription. Special proteins called transcription factors, which bind to promoters, subsequently provide binding sites for the RNA polymerase, which is the enzyme that transcribes the RNA. Promoters in the Archaea and Eukaryota domains often contain a nucleotide sequence TATA, which is called a TATA box. The TATA box is usually 25 nucleotides upstream of the transcription start site, and it is the location at which the DNA is unwound.

Enhancers

Enhancers are DNA sequences that regulate gene expression by providing a binding site for proteins that regulate RNA polymerase in transcribing proteins. Enhancers can greatly increase the expression of genes in their range. They can be hundreds or thousands of base pairs upstream or downstream from the genes they control. Some enhancers are located within the gene they control. Enhancers are functional over large distances. Most genes are controlled by two or three enhancers, but some may be controlled by more. Enhancers provide bonding sites for regulatory proteins that either promotes or inhibits RNA polymerase activity. Enhancers are more common in eukaryotes than prokaryotes.

Transcription factors

Transcription factors are proteins that help regulate gene expression in eukaryotes and prokaryotes. Transcription factors bind to the DNA and determine if that sequence of DNA is transcribed into mRNA and then into proteins. In eukaryotes that have promoter or enhancer regions, transcription factors bind near these regions and increase the ability of the RNA polymerase to start transcription. In eukaryotes, because genes are typically turned "off," transcription factors typically work to turn genes "on." The opposite is often true in bacteria, and transcription factors often work to turn genes off.

Operons

Operons control gene regulation. Specifically, operons allow cells to only code for proteins as they are needed by the cell. This allows cells to conserve energy. Operons are segments of DNA or groups of genes that are controlled by one promoter. Operons consist of an operator, a promoter, and structural gene(s). The operator provides a binding side for a repressor that inhibits the binding of RNA polymerase. The promoter provides the binding site for the RNA polymerase. The structural genes provide the sequence that codes for a protein. Operons are transcribed as single units and code for a single mRNA molecule, which produces proteins with related functions. Operons have been found in prokaryotes, eukaryotes, and viruses. For example, the lac (lactose) operon in certain bacteria controls the production of the enzymes needed to digest any lactose in the cell. If lactose is already available in the cell, the lactose binds to the repressor protein to prevent the repressor protein from binding to the operator. The gene is transcribed, and the enzymes necessary for the digestion of the lactose are produced. If there is no lactose that needs to be digested, the repressor protein binds to the operator. The gene is not transcribed, and the enzyme is not produced.

Epigenetics

Epigenetics is the study of modifications in specific gene expression caused by factors that are not genetic. These factors do not cause alternations in the cell's DNA. Epigenetics studies factors or mechanisms that determine if genes are active (switched on) or dormant (switched off). These mechanisms can alter gene functions or gene expressions without altering the sequences of the DNA itself. Modifications to the proteins such as histones that are associated with DNA can switch genes on or off. Some modifications determine the activity level of a gene, which in turn affects a physiological aspect of the health of an individual. The main type of modification is the addition of a methyl group to the histones, known as methylation. Acetylation, the addition of an acetyl group, and phosphorylation, the addition of a phosphoryl group, are also modifications to the proteins associated with DNA that can switch genes on or off or affect their activity level.

Differential gene expression

Because every cell in an organism has an identical genome, the DNA molecules of every cell of that organism are identical. Cells must be specialized for their specific roles. For example, in mammals, there are numerous types of cells such as epithelial cells, nerve cells, blood cells, liver cells, fat cells, and bone cells. The various types of cells differentiate through differential gene expression. Differential gene expression is the expression of different sets of gene by cells with identical DNA molecules. The unused genes in a differentiated cell remain in the cell; they just are not expressed. Actually, only a few genes are expressed in each cell. For example, during mammalian embryonic development, the undifferentiated zygote undergoes cell division through mitosis. As the number of cells increases, selected cells undergo differentiation to become specialized components in the developing tissues of the embryo.

Stem cells

Stem cells are undifferentiated cells that can divide without limit and that can differentiate to produce the specialized cells that each organism needs. Stem cells have varying degrees of potency. Stem cells can be pluripotent or multipotent depending on their source. Embryonic stem cells are harvested from the embryo at the blastocyst stage or from the developing gonads of the embryo. Early embryonic stem cells are pluripotent. This means they have not undergone any differentiation and have the ability to become any special type of cell. After embryonic stem cells begin to differentiate, they may be limited to specializing into a specific tissue type. These stem cells are considered to be multipotent because they can only develop into a few different types of cells. Adult stem cells, also called somatic stem cells, are harvested from organs and tissues and can differentiate into those types of cells in that particular organ or tissue. Umbilical cord blood stem cells can be harvested from the umbilical cord of a newborn baby. Adult stem cells and umbilical cord blood stem cells are multipotent. Induced pluripotent cells (iPS) are somatic cells that have been manipulated to act like pluripotent cells. Experiments have shown that iPS may be useful in treating diseases.

Mutations and mutagens

Mutations are errors in DNA replication. Mutagens are physical and chemical agents that cause these changes or errors in DNA replication. Mutagens are external factors to an organism. The first mutagens discovered were carcinogens or cancer-causing substances. Other mutagens include ionizing radiation such as ultraviolet radiation, x-rays, and gamma radiation. Viruses and microorganisms that integrate into chromosomes and switch genes on or off causing cancer are mutagens. Mutagens include environmental poisons such as asbestos, coal tars, tobacco, and benzene. Alcohol and diets high in fat have been shown to be mutagenic. Not all mutations are caused by mutagens. Spontaneous mutations can occur in DNA due to molecular decay. Spontaneous errors in DNA replication, repair, and recombination can also cause mutations.

Mutations are changes in DNA sequences. Point mutations are changes in a single nucleotide or at one "point" in a DNA sequence. Three types of point mutations include missense, silent, and nonsense. Missense mutations code for the wrong protein. Silent mutations do not change the function of the protein. Nonsense mutations stop protein synthesis early, resulting in no functioning protein. Deletions and insertions remove and add one or more nucleotides to the DNA sequence, which can remove or add amino acids to the protein, changing the function. Deletions and insertions can also cause a frameshift mutation in which the nucleotides are grouped incorrectly in sets of three. Mutations can also occur on the chromosomal level. For example, an inversion is when a piece of the chromosome inverts or flips its orientation. Mutations can occur in somatic (body) cells and germ cells (egg and sperm) at any time in an organism's life. Somatic mutations develop after conception and occur in an organism's body cells such as bone cells, liver cells, or brain cells. Somatic mutations cannot be passed on from parent to offspring. The mutation is limited to the specific descendent of the cell in which the mutation occurred. The mutation is not in the other body cells unless they are descendants of the originally mutated cell. Somatic mutations may cause cancer or diseases. Some somatic mutations are silent. Germline mutations are present at conception and occur in an organism's germ cells, which are only egg and sperms cells. Germline mutations may be passed on from parent to offspring. Germline mutations will be present in every cell of an offspring that inherits a germline mutation. Germline mutations may cause diseases. Some germline mutations are silent.

Gel electrophoresis

Gel electrophoresis is a technique used to separate macromolecules such as nucleic acids and proteins. Fragments of DNA and RNA are separated according to length. Proteins are separated according to length and charge. The technique is relatively simple. For example, to separate DNA strands, a solution containing the DNA strand is placed in a gel. When an electric current is passed through the gel, the DNA strands migrate from the negative end of the container to the positive end due to their negative charge because of their phosphate ions. Shorter DNA strands migrate faster than the longer DNA strands. This results in a series of bands. Each band contains DNA strands of a specific length. A DNA standard is placed in the gel to provide a reference to determine the strand length. Lengths are measured in base pairs (bps).

Microscopy

Microscopy is used in microbiology. Bacteria, viruses, cell components, and molecules are too small to be seen by the naked eye. Several types of microscopes are available to examine these samples. There are light microscopes, which use visible light to study samples, and electron microscopes, which use beams of electrons. The light microscope (also called the compound microscope) uses two types of lenses (ocular and objective) to magnify objects. These are typically used when studying samples at the cellular level. Basic compound light microscopes are typically used in high school biology classes. Other compound light microscopes such as the dark-field microscope, phase-contrast microscope, and the fluorescent microscope are available for more specific uses. For tiny samples, such as viruses, cell components, or individual molecules, electron microscopes can be used. Electron microscopes use beams of electrons instead of light. Because beams of electrons have shorter wavelengths, electron microscopes have greater resolution than light microscopes. Resolution is the ability of a lens to reveal two points as being distinct. The two types of electron microscopy are transmission electron microscopy (TEM) and scanning electron microscopy (SEM). SEM is a newer technology than TEM and produces three-dimensional images.

Polymerase chain reaction (PCR)

The polymerase chain reaction (PCR) is a laboratory technique used to rapidly copy selected segments of DNA from DNA molecules without cloning. PCR requires only a single cell such as from sperm, hair, or blood to obtain the targeted DNA. PCR is a hot-and-cold cycled reaction that uses a special heat-tolerant polymerase that has been extracted from bacteria. The DNA sample is combined with this special DNA polymerase, primers, and free nucleotides. Primers are synthetic strands of DNA containing just a few bases. Primers attach to the ends of the targeted DNA sequence and act as the substrate for the polymerase. At high temperatures, the DNA molecules separate into two strands, and each strand unwinds. Then the mixture is cooled, and the primers bind to the ends of the targeted DNA segment. The polymerase initiates synthesis between the two primers. When the temperature cycles up again, the DNA separates again into two strands, and the cycle repeats. After 30 cycles, which takes less than three hours, there are more than half a billion of the needed targeted DNA segments.

DNA sequencing

DNA sequencing is a laboratory technique used to determine the order or linear sequence of nucleotides of DNA fragments. A polymerase chain reaction (PCR) is used to isolate the needed DNA segment or DNA template. During PCR, some of each of the nucleotides containing the four bases, G, C, A, and T, is chemically altered and fluorescently tagged with different colors of dye. Also, the chemically altered nucleotides have the dideoxyribose sugar, which contains one less oxygen atom than the usual deoxyribose. When synthesis begins, the polymerase randomly adds either a regular nucleotide or an altered nucleotide. If the polymerase adds an altered nucleotide, synthesis stops. This way, each DNA fragment of the same length is tagged with the same color. Then, electrophoresis is used to separate DNA fragments according to length. The DNA sequence can be read by reading the tags of the shortest fragments to the tags of the longest fragments.

Human Genome Project

In 1990, the Human Genome Project (HGP), which involved scientists from 16 laboratories located in at least 6 different countries, was launched to map the human genome. The project was completed in 2003. The human genome consists of approximately 3.12 billion paired nucleotides. The results were surprising. Prior to the project, scientists thought that the human genome would consist of approximately 100,000 to 140,000 genes, but research showed that the human genome consists of only about 21,000 genes. Prior to the project, scientists thought that each gene coded for one specific protein, but with only 21,000 genes, this could not be correct. Genes must be able to code for more than one protein. Furthermore, scientists discovered that only about 1% of the genes actually code for proteins. Originally, the noncoding DNA was actually called junk DNA, but new research has shown that these genes are involved in gene expressions and in turning genes on and off. Today, more than 21,000 genes have been identified. The genomes of several plants, animal, fungi, protists, bacteria, viruses, and even cell organelles have been studied and mapped. Interesting comparisons can be made between these genomes. For example, the number of genes in an organism's genome does not indicate the complexity of that organism. Humans have approximately 21,000 genes, but the simpler roundworms have approximately 26,000 genes.

Gene therapy

Gene therapy is an experimental but promising technique that introduces new genes into an organism to correct a specific disease caused by a defective gene. In gene therapy, the defective gene is replaced by a properly functioning gene. Gene therapy is most promising for diseases that are caused by a single defective gene. For example, gene therapy was first successfully used to treat severe combined immunodeficiency (SCID). One type of SCID is caused by a single defective gene on the X chromosome. Doctors removed some bone marrow from the test subjects, injected a retrovirus that was carrying the gene, and then reimplanted the bone marrow. The bone marrow cells then have the correct DNA sequence for the production of proteins for much-needed enzymes. Unfortunately, some of the first recipients developed leukemia, and the trials were halted. Later, researchers discovered that the leukemias were related to the location of the insertion of the retroviral vectors.

Cloning

Clones are exact biological copies of genes, cells, or multicellular organisms. There are natural clones and artificial clones. Many clones are produced in nature. Animals that can reproduce asexually by fragmentation or budding produce natural clones. Some plants such as strawberries can reproduce by stolons. Typically, in biology, cloning refers to gene cloning or the cloning of organisms. Gene cloning is the process of splicing genes that are needed to code for a specific protein and introducing them into a new cell with a DNA vector. Gene cloning has been used with bacteria in the production of human insulin and a human growth hormone replacement. Cloning can also occur with an entire organism. This type of cloning is called a somatic cell nuclear transfer. The first mammal clone was Dolly the sheep. In this procedure, a nucleus of a somatic cell from the sheep to be cloned was transferred or injected into a denucleated egg cell of the surrogate mother sheep. The egg was stimulated to divide by electric shock, and then the embryo was implanted into the uterus of the surrogate mother. Dolly was born identical to the egg nucleus donor, not the surrogate mother. Dolly and other cloned mammals typically have serious health problems. Dolly aged prematurely possibly due to the shortened telomeres from the adult somatic cell nucleus.

Genetically engineered cells

Genetic engineering is the manipulation of DNA outside of normal reproduction. This modified DNA is called recombinant DNA. Genetic engineering is prevalent in gene cloning, which is used in the production of genetically modified (GM) organisms and the production of GM food. Gene cloning involves cloning a specific gene that is needed for a specific purpose. Genes can be inserted into cells of an entirely different species. Genetically engineered cells are also called transgenic cells. GM organisms such plants or crops contain recombinant DNA. Many types of organisms such as plants, animals, fungi, and bacteria have been genetically modified. GM crops such as corn and soybeans can be engineered to be herbicide resistant to ensure that herbicides kill the weeds but not the crop plants. Crops can also be modified to be pest resistant in order to kill the insects that might damage the crops. Also, several foods can be genetically modified to increase the nutritional value.

Interactions

Biosphere

The components of the biosphere from smallest to largest are organisms, populations, communities, ecosystems, and biomes. Living things are called organisms. Organisms of the same species make up a population. All of the populations in an area make up the community. The community combined with the physical environment for a region form an ecosystem. Several ecosystems are grouped together to form large geographic regions called biomes. The biosphere is the worldwide ecosystem and is the sum of all the world's biomes.

The biosphere consists of numerous biomes. A biome is a large region that supports a specific community. Each biome has a characteristic climate and geography. Differences in latitude, altitude, and worldwide patterns affect temperature, precipitation, and humidity. Biomes can be classified as terrestrial or aquatic biomes. Terrestrial biomes include ecosystems with land environments. Terrestrial biomes include the tundra, coniferous forest, temperate broadleaf forest, temperate grassland, chaparral, desert, savannas, and tropical forests. Terrestrial biomes tend to grade into each other in regions called ecotones. Aquatic biomes are water-dwelling ecosystems. Aquatic biomes include lakes, coral reefs, rivers, oceanic pelagic zone, estuaries, intertidal zone, and the abyssal zone.

Ecosystems

An ecosystem is the basic unit of ecology. An ecosystem is the sum of all the biotic and abiotic factors in an area. Biotic factors are all living things such as plant, animals, fungi, and microorganisms. Abiotic factors include the light, water, air, temperature, and soil in an area. Ecosystems obtain the energy they need from sunlight. Ecosystems contain biogeochemical cycles such as the hydrologic cycle and the nitrogen cycle. Ecosystems are generally classified as either terrestrial or aquatic. All of the living things within an ecosystem are called its community. The number and variety of its community describes the ecosystem's biodiversity. Each ecosystem can only support a limited number of organisms known as the carrying capacity.

Every ecosystem consists of multiple abiotic and biotic factors. Abiotic factors are the nonbiological physical and chemical factors that affect the ecosystem. Abiotic factors include soil type, atmospheric conditions, sunlight, water, chemical elements such as acidity in the soil, wind, and natural disturbances such as forest fires. In aquatic ecosystems, abiotic factors include salinity, turbidity, water depth, current, temperature, and light. Biotic factors are all of the living organisms in the ecosystem. Biotic factors include plants, algae, fungi, bacteria, archaea, animals, and protozoa.

Community

A community is all of the populations of different species that live in an area and interact with each other. Community interaction can be intraspecific or interspecific. Intraspecific interactions occur between members of the same species. Interspecific interactions occur between members of different species. Different types of interactions include competition, predation, and symbiosis. Communities with high diversity are more complex and more stable than communities with low diversity. The level of diversity can be seen in a food web of the community, which shows all the feeding relationships within the community.

Population

A population is a group of all the individuals of one species in a specific area or region at a certain time. A species is a group of organisms than can breed and produce fertile offspring. There may be many populations of a specific species in a large geographic region. Ecologists study the size, density, and growth rate of populations to determine their stability. The population density is the number of individuals per unit of area. Growth rates may be exponential or logistic. Population size continuously changes with births, deaths, and migrations. Ecologists also study the dispersion or how the individuals are spaced within a population. Some species form clusters. Others are evenly or randomly spaced. Every population has limiting factors. Changes in the environment can reduce population size. Geography can limit population size. The individuals of a population react with each other and with other organisms in the community. Competition and predation affect population size. All the populations in an ecosystem make up the community of that ecosystem.

Population size

Population size is affected by resource availability and abiotic factors. As the population density increases, intraspecific competition for available resources intensifies. If the availability of resources decreases, death rates may increase and birth rates may decrease. For example, territoriality for mating or nesting may limit available resources for individuals in a population and limit the population size. Abiotic factors such as temperature, rainfall, wind, and light intensity all influence the population size. For example, temperatures near a species' tolerance limit may decrease the population. Natural disasters such as fire or flood can destroy resources and greatly decrease a population's size. In general, any abiotic factor that reduces or limits resources will also reduce or limit population size.

Habitat and niche

The habitat of an organism is the type of place where an organism usually lives. A habitat is a piece of an environmental area. A habitat may be a geographic area or even the body of another organism. The habitat describes an organism's natural living environment. A habitat includes biotic and abiotic factors such as temperature, light, food resources, and predators. Whereas a habitat describes an organism's "home," a niche can be thought of as an organism's "occupation." A niche describes an organism's functional role in the community. A niche can be quite complex because it should include the impacts that the organism has on the biotic and abiotic surroundings. Niches can be broad or narrow. The niche describes the way an organism uses its habitat.

Competition and predation

Feeding relationships between organisms can affect population size. Competition and predation both tend to limit population size. Competition occurs when two individuals need the same resource. Predation occurs when one individual is the resource for another individual. Competition occurs when individuals share a resource in the habitat. This competition can be intraspecific, which is between members of the same species, or interspecific, which is between members of different species. Intraspecific competition reduces resources as that species' own population increases. This limits population growth. Interspecific competition reduces resources as a different species uses those same resources. Predation occurs when one species is a food resource for another species. Predator and prey populations can cycle over a range of years. If prey resources increase, predator numbers increase. An example of the predator-prey population cycle is the Canadian lynx and snowshoe hare.

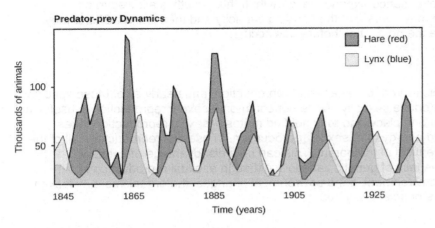

Logistic population growth model

Populations vary over time due to deaths, births, immigration, and emigration. In most situations, resources such as food, water, and shelter are limited. Each environment or habitat can only support a limited number of individuals. This is known as the carrying capacity. Population growth models that factor in the carrying capacity are called logistic growth models. With logistic population growth models, the rate of population growth decreases as the population size increases. Logistic growth graphs as an S-shaped curve. Comparing logistic growth and exponential growth shows that the graph for exponential growth continues to become steeper, but the graph for logistics growth levels off once the population reaches the carrying capacity. As the population increases, fewer resources are available per individual. This limits the number of individuals that can occupy that environment or habitat.

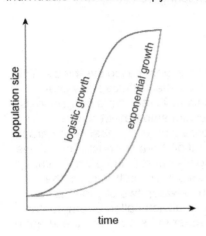

Exponential population growth model

Populations change over time due to births, deaths, and migrations. Sometimes, conditions are near ideal and populations can increase at their maximum rate exhibiting exponential growth. Exponential growth is growth in which the rate of change is proportional to the increasing size in an exponential progression. Exponential growth graphs as a J-shaped curve. Exponential growth is often observed in single-celled organisms such as bacteria or protozoa in which the population or number of cells increases by a factor of two per unit of time. One cell divides into two cells, which divide into four cells, and so forth. The exponential growth model describes population growth under ideal conditions. It does not take limiting factors or carrying capacity into account. Realistically, exponential growth cannot occur indefinitely, but it may occur for a period of time. It does show a species' capacity for increase and may be helpful when studying a particular species or ecosystem. For example, if a species with no natural predator is introduced into a new habitat, that species may experience exponential growth. If this growth is allowed to go unchecked, the population may overshoot the carrying capacity and then starve. Efforts may need to be taken to reduce the population before this occurs.

Asexual reproduction

Very few species of animals reproduce by asexual reproduction, and nearly all of those species also have the ability to reproduce sexually. While not common, asexual reproduction is useful for animals that tend to stay in one place and may not find mates. Asexual reproduction takes considerably less effort and energy than sexual reproduction. In asexual reproduction, all of the offspring are genetically identical to the parent. This can be a disadvantage of asexual reproduction because of the lack of genetic variation. Although asexual reproduction is advantageous in a stable environment, if the environment changes, the organisms may lack the genetic variability to survive or selectively adapt.

Life histories

The life history of a species describes the typical organism's life specifically concerning its survival and reproduction from birth through reproduction to death. Life histories can typically be classified as opportunistic life histories or equilibrial life histories. Species exhibiting opportunistic life histories are typically small, short-lived organisms that have a high reproductive capacity but invest little time and care into their offspring. Their population sizes tend to oscillate significantly over periods of several years. Species exhibiting equilibrial life histories are typically large, long-lived organisms that have a low reproductive capacity but invest much time and care into their offspring. Their populations tend to fluctuate within a smaller range. A general observation is that species that tend to produce numerous offspring typically tend to invest little care into that offspring, resulting in a high mortality rate of that offspring. Organisms of species that tend to produce few offspring typically invest much more care into that offspring, resulting in a lower mortality rate.

Symbiosis

Many species share a special nutritional relationship with another species, called symbiosis. The term symbiosis means "living together." In symbiosis, two organisms share a close physical relationship that can be helpful, harmful, or neutral for each organism. Three forms of symbiosis are parasitism, commensalism, and mutualism. Parasitism is a relationship between two organisms in which one organism is the parasite, and the other organism is the host. The parasite benefits from the relationship because the parasite obtains its nutrition from the host. The host is harmed from the relationship because the parasite is using the host's energy and giving nothing in return. For example, a tick and a dog share a parasitic relationship in which the tick is the parasite, and the dog is the host. Commensalism is a relationship between two organisms in which one benefits, and the other is not affected. For example, a small fish called a remora can attach to the belly of a shark and ride along. The remora is safe under the shark, and the shark is not affected. Mutualism is a relationship between two organisms in which both organisms benefit. For example, a rhinoceros usually can be seen with a few tick birds perched on its back. The tick birds are helped by the easy food source of ticks, and the rhino benefits from the tick removal.

Predation

Predation is a special nutritional relationship in which one organism is the predator, and the other organism is the prey. The predator benefits from the relationship, but the prey is harmed. The predator hunts and kills the prey for food. The predator is specially adapted to hunt its prey, and the prey is specially adapted to escape its predator. While predators harm (kill) their individual prey, predation usually helps the prey species. Predation keeps the population of the prey species under control and prevents them from overshooting the carrying capacity, which often leads to starvation. Also, predation usually helps to remove weak or slow members of the prey species leaving the healthier, stronger, and better adapted individuals to reproduce. Examples of predator-prey relationships include lions and zebras, snakes and rats, and hawks and rabbits.

Competition and territoriality

Competition is a relationship between two organisms in which the organisms compete for the same vital resource that is in short supply. Typically, both organisms are harmed, but one is usually harmed more than the other. They could be competing for resources such as food, water, mates, and space. Interspecific competition is between members of different species. Intraspecific competition is between members of the same species. Competition provides an avenue for natural selection. Territoriality can be considered to be a type of interspecific competition for space. Many animals including mammals, birds, reptiles, fish, spiders, and insects have exhibited territorial behavior. Once territories are established, there are fewer conflicts between organisms. For example, a male redwing blackbird can establish a large territory. By singing and flashing his red patches, he is able to warn other males to avoid his territory, and they can avoid fighting.

Altruistic behaviors

Altruism is a self-sacrificing behavior in which an individual animal may serve or protect another animal. Altruism is seen in the social insects honey bees and fire ants. For example, in a honey bee colony, there is one queen, many workers (females), and drones (males) only during the mating seasons. Adult workers do all the work of the hive and will die defending it. Another example of altruism is seen in a naked mole rat colony. Each colony has one queen that mates with a few males, and the rest of the colony is nonbreeding and lives to service the queen, her mates, and her offspring.

Primary and secondary succession

Ecological succession is the process by which climax communities come into existence or are replaced by new climax communities when they are greatly changed or destroyed. The two types of ecological succession are primary succession and secondary succession. Primary succession occurs in a region where there is no soil and that has never been populated such as a new volcanic island or a region where a glacier has retreated. During the pioneer stage, the progression of species is typically lichen and algae, followed by small annual plants, then perennial herbs and grasses. During the intermediate stage, shrubs, grasses, and shade-intolerant trees are dominant. Finally, after hundreds of years, a climax community is reached with shade-tolerant trees. Secondary succession occurs when a climax community is destroyed or nearly destroyed such as after a forest fire or in an abandoned field. With secondary succession, the area starts with soil and seeds from the original climax community. Typically, in the first two years, weeds and annuals are dominant. This is followed by grasses and biennials. In a few years, shrubs and perennials are dominant followed by pine trees, which are eventually replaced by deciduous trees. Secondary succession takes place in less than 100 years.

Terrestrial biomes

Terrestrial biomes are classified predominantly by their vegetation, which is primarily determined by precipitation and rainfall. Tropical rainforests experience the highest annual precipitation and relatively high temperatures. The dominant vegetation in tropical forests is tall evergreen trees. Temperate deciduous forests experience moderate precipitation and temperatures. The dominant vegetation is deciduous trees. Boreal forests experience moderate precipitation and lower temperatures. The dominant vegetation is coniferous trees. The tundra experiences lower precipitation and cold temperatures. The dominant vegetation is shrubs. The savanna experiences lower precipitation and high temperatures. The dominant vegetation is grasses. Deserts experience the lowest precipitation and the hottest temperatures. The dominant vegetation is scattered thorny plants.

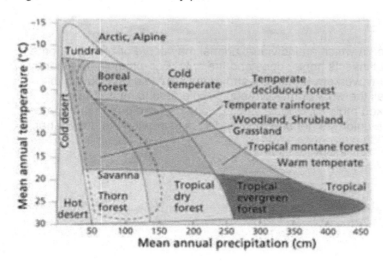

Aquatic biomes

Aquatic biomes are characterized by multiple factors including the temperature of the water, the amount of dissolved solids in the water, the availability of light, the depth of the water, and the material at the bottom of the biome. Aquatic biomes are divided the two categories of marine regions and fresh-water regions based on the amount of dissolved salt in the water. Marine biomes include the pelagic zone, the benthic zone, coral reefs, and estuaries. Freshwater biomes include lakes, ponds, rivers, and streams. Marine biomes have a salinity of at least 35 parts per thousand. Freshwater biomes have a salinity that is less than 0.5 parts per thousand. Freshwater biomes include lakes, ponds, streams, and rivers. Lakes and ponds, which are relatively stationary, consist of two zones: the littoral zone and the limnetic zone. The littoral zone is closest to the shore and is home to many plants both floating and rooted, invertebrates, crustaceans, amphibians, and fish. The limnetic zone is further from the shore and has no rooted plants. Rivers and streams typically originate in the mountains and make their way to the oceans. Because this water is running and colder, it contains different plants and animals than lakes and ponds. Salmon, trout, crayfish, plants, and algae are found in rivers and streams.

Marine biomes

Marine regions are located in three broad areas: the ocean, estuaries, and coral reefs. The pelagic zone is out in the open ocean. Organisms in the pelagic zone include phytoplankton such as algae and bacteria; zooplankton such as protozoa and crustaceans; and larger animals such as squid, sharks, and whales. The benthic zone is located underneath the pelagic zone. The type of community in each region is largely determined by whether the bottom is sandy, muddy, or rocky. Organisms in the benthic zone can include sponges, clams, oysters, starfish, sea anemones, sea urchins, worms, and fish. The deepest part of the benthic zone is called the abyssal plain. This is the deep ocean floor, which is home to numerous scavengers, many of which have light-generating capability. Coral reefs are located in warm, shallow water. Corals are small colonial animals that share a mutualistic relationship with algae. Estuaries are somewhat-enclosed coastal regions where water from rivers and streams is mixed with seawater.

Energy flow

Energy flow through an ecosystem can be tracked through an energy pyramid. An energy pyramid shows how energy is transferred from one trophic level to another. Producers always form the base of an energy pyramid, and the consumers form successive levels above the producers. Producers only store about 1% of the solar energy they receive. Then, each successive level only uses 10% of the energy of the previous level. In this energy pyramid, grass, which is a producer, uses 1,000 kcal of energy. Then the grasshoppers, which are primary consumers, use 10% of that 1,000 kcal or 100 kcal. Next, the moles, which are secondary consumers, use 10% of that 100 kcal or 10 kcal. Finally, the owl, which is a tertiary consumer, uses 10% of that 10 kcal or 1 kcal of energy.

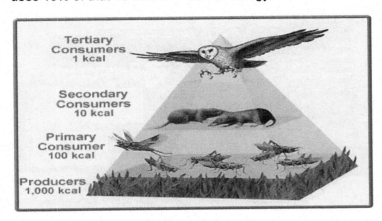

Energy flow through an ecosystem can be illustrated by a food web. Energy moves through the food web in the direction of the arrows. In the food web below, producers such as grass, trees, and shrubs use energy from the sun to produce food through photosynthesis. Herbivores or primary consumers such as squirrels, grasshoppers, and rabbits obtain energy by eating the producers. Secondary consumers, which are carnivores such as snakes and shrews, obtain energy by eating the primary consumers. Tertiary consumers, which are carnivores such as hawks and mountain lions, obtain energy by eating the secondary consumers. Note that the hawk and the mountain lion can also be considered quaternary consumers in this food web if a different food chain within the web is followed.

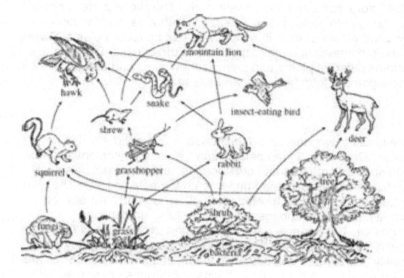

Water cycle

The water cycle, also referred to as the hydrologic cycle, is a biogeochemical cycle that describes the continuous movement of the Earth's water. Water in the form of precipitation such as rain or snow moves from the atmosphere to the ground. The water is collected in oceans, lakes, rivers, and other bodies of water. Heat from the sun causes water to evaporate from oceans, lakes, rivers, and other bodies of water. As plants transpire, this water also undergoes evaporation. This water vapor collects in the sky and forms clouds. As the water vapor in the clouds cools, the water vapor condenses or sublimes depending on the conditions. Then, water moves back to the ground in the form of precipitation.

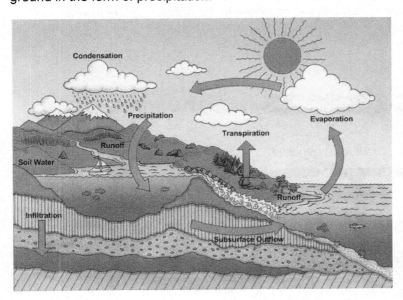

Carbon cycle

The carbon cycle is a biogeochemical cycle that describes the continuous movement of the Earth's carbon. Carbon is in the atmosphere, the soil, living organisms, fossil fuels, oceans, and freshwater systems. These areas are referred to as carbon reservoirs. Carbon flows between these reservoirs in an exchange called the carbon cycle. In the atmosphere, carbon is in the form of carbon dioxide. Carbon moves from the atmosphere to plants through the process of photosynthesis. Carbon moves from plants to animals through food chains. Carbon moves from living organisms to the soil when these organisms die. Carbon moves from living organisms to the atmosphere through cellular respiration. Carbon moves from fossil fuels to the atmosphere when fossil fuels are burned. Carbon moves from the atmosphere to the oceans and freshwater systems through absorption.

The carbon cycle

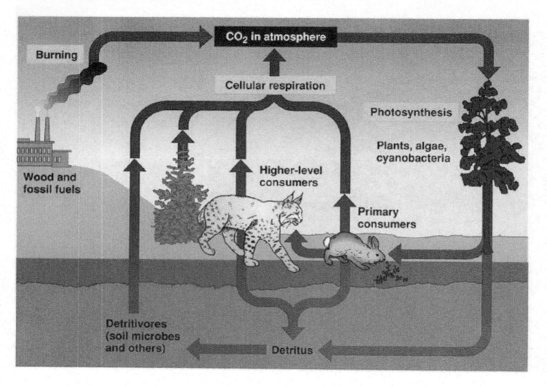

Nitrogen cycle

The nitrogen cycle is a biogeochemical cycle that describes the continuous movement of the Earth's nitrogen. Approximately 78% of the Earth's atmosphere consists of nitrogen in its elemental form N_2. Nitrogen is essential to the formation of proteins, but most organisms cannot use nitrogen in this form and require the nitrogen to be converted into some form of nitrates. Lightning can cause nitrates to form in the atmosphere, which can be carried to the soil by rain to be used by plants. Legumes have nitrogen-fixing bacteria in their roots, which can convert the N_2 to ammonia (NH_3). Nitrifying bacteria in the soil can also convert ammonia into nitrates. Plants absorb nitrates from the soil, and animals can consume the plants and other animals for protein. Denitrifying bacteria can convert unused nitrates back to nitrogen to be returned to the atmosphere.

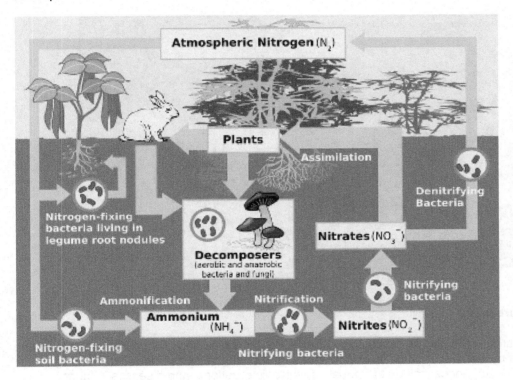

Phosphorus cycle

The phosphorus cycle is a biogeochemical cycle that describes the continuous movement of the Earth's phosphorus. Phosphorus is found in rocks. When these rocks weather and erode, the phosphorus moves into the soil. The phosphorus found in the soil and rocks is in the form of phosphates or compounds with the PO_4^{3-} ion. When it rains, phosphates can be dissolved into the water. Plants are able to use phosphates from the soil. Plants need phosphorus for growth and development. Phosphorus is also a component of DNA, RNA, ATP, cell membranes, and bones. Plants and algae can absorb phosphate ions from the water and convert them into many organic compounds. Animals can get phosphorus from eating plants or other animals and drinking water. When organisms die, the phosphorus is returned to the soil. This is the slowest of all biogeochemical cycles.

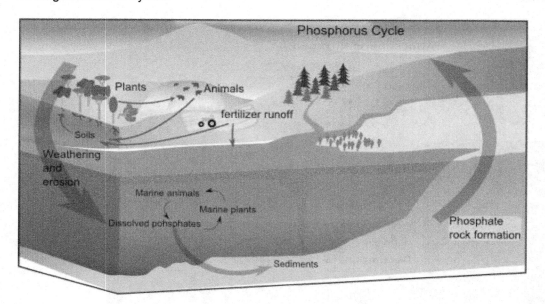

Natural disturbances

A natural disturbance is an event caused by nature, not human activity. Natural disturbances can be brought on by weather such as fires from lightning, droughts, storms, wind, and freezing. Other natural disturbances include earthquakes, volcanic eruptions, and diseases. Natural disturbances can disrupt or disturb the ecosystem in many ways such as altering resources or removing individuals from the community. Natural disturbances can cover small regions, or they can affect an entire ecosystem. The effect may be long lasting and take several years to recover, or the effect may be minor and take only a few months to recover.

Fragmentation

Fragmentation of ecosystems decreases biodiversity. Habitats can become fragmented due to natural disturbances such as fire, volcanic activity, and climate change. Fragmented habitats have been segmented into patches each with its own periphery. The resulting habitats are basically isolated from each other reproductively limiting genetic variation. Limited genetic variation can lead to extinction of that population resulting in lower biodiversity. Some of the original habitat is destroyed during fragmentation. Plants and nonmotile animals can be destroyed. Motile animals can move into the other fragments, which can lead to overcrowding. Habitat fragmentation leads to populations being split into smaller groups. The number of species in a fragment is determined by the area of the fragment. Fragments with small areas have small populations. Small populations may become endangered or extinct due to the loss of genetic variation. Small fluctuations in resources or climate can be catastrophic in small populations.

Larger populations may be able to overcome these fluctuations. Habitat fragmentation reduces the total area of the habitat. The resulting habitats have less interior and more periphery. The resulting habitats may not have enough food or other resource necessary to support a species. Species living near the periphery are more susceptible to predators, wind, fires, and other limiting factors.

Human population

Human population has been increasing at a near-exponential rate for the past 50 years. As the human population increases, the demand for resources such as food, water, land, and energy also increases. As the human population increases, species decrease largely due to habitat destruction, introduced species, and overhunting. The increased greenhouse gases and resulting climate changes have also significantly affected many ecosystems as temperatures rise and habitats are slowly changed or even destroyed. Increasing human population means increasing pollution, which harms habitats. Many animals have become extinct due to the effects of an exponentially increasing human population. High rates of extinction greatly reduce biodiversity.

Habitat destruction by humans

Many habitats have been altered or destroyed by humans. In fact, habitat destruction brought about by human endeavors has been the most significant cause of species extinctions resulting in the decrease in biodiversity throughout the world. As the human population has increased exponentially, the extinction rate has also increased exponentially. This is largely due to habitat destruction by humans. Humans use many resources in their various enterprises including agriculture, industry, mining, logging, and recreation. Humans have cleared much land for agriculture and urban developments. As habitats are destroyed, species are either destroyed or displaced. Often, habitats are fragmented into smaller areas, which only allow for small populations that are under threat by predators, diseases, weather, and limited resources. Especially hard hit are areas near the coastline, estuaries, and coral reefs. Nearly half of all mangrove ecosystems have been destroyed by human activity. Coral reefs have nearly been decimated from pollution such as oil spills and exploitation from the aquarium fish market and coral market.

Introduced species

Introduced species are species that are moved into new geographic regions by humans. The introduction can be intentional such as the introduction of livestock including cattle, pigs, and goats to the United States or unintentional such as the introduction of Dutch elm disease, which has damaged and killed thousands of American elms. Introduced species are also called invasive or nonnative species. Introduced species typically cause a decrease in biodiversity. Introduced species can disrupt their new communities by using limited resources and preying on other members of the community. Introduced species are often free from predators and can reproduce exponentially. Introduced species are contributors or even responsible for numerous extinctions. For example, zebra mussels, which are native to the Black Sea and the Caspian Sea, were accidentally introduced to the Great Lakes. The zebra mussels greatly reduced the amount of plankton available for the native mussel species, many of which are now endangered.

Remediation

In recent years, much work has been done in remediation or land rehabilitation especially in the areas of reforestation and mine reclamation. In remediation, land is returned at least to some degree to its former state. Mining reclamation incudes the backfilling of open-pit mines and covering ores containing sulfides to prevent rain from mixing with them to produce sulfuric acid. Reforestation is the restocking of forest and wetlands. Restoration can at least partially help offset some of the damaging effects brought about by the deforestation. Reforestation can help reduce global warming due to an increase in the absorption of light by the trees. Restoration can at least partially help to restore the carbon cycle and lessen greenhouse gases. Reforestation also helps to counter erosion. Reforestation can help maintain or preserve the biodiversity of the region and possibly increase biodiversity if new organisms immigrate into the region.

Plant cells and animal cells

Plant cells and animal cells both have a nucleus, cytoplasm, cell membrane, ribosomes, mitochondria, endoplasmic reticulum, Golgi apparatus, and vacuoles. Plant cells have only one or two extremely large vacuoles. Animal cells typically have several small vacuoles. Plant cells have chloroplasts for photosynthesis because plants are autotrophs. Animal cells do not have chloroplasts because they are heterotrophs. Plant cells have a rectangular shape due to the cell wall, and animal cells have more of a circular shape. Animal cells have centrioles, but only some plant cells have centrioles.

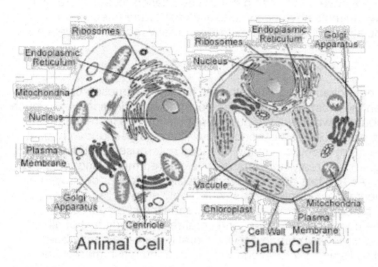

Cell membranes

The cell membrane, also referred to as the plasma membrane, is a thin semipermeable membrane of lipids and proteins. The cell membrane isolates the cell from its external environment while still enabling the cell to communicate with that outside environment. It consists of a phospholipid bilayer, or double layer, with the hydrophilic ends of the outer layer facing the external environment, the inner layer facing the inside of the cell, and the hydrophobic ends facing each other. Cholesterol in the cell membrane adds stiffness and flexibility. Glycolipids help the cell to recognize other cells of the organisms. The proteins in the cell membrane help give the cells shape. Special proteins help the cell communicate with its external environment. Other proteins transport molecules across the cell membrane.

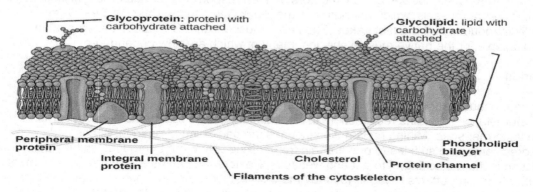

Nucleus

Typically, a eukaryote has only one nucleus that takes up approximately 10% of the volume of the cell. Components of the nucleus include the nuclear envelope, nucleoplasm, chromatin, and nucleolus. The nuclear envelope is a double-layer membrane with the outer layer connected to the endoplasmic reticulum. The nucleus can communicate with the rest of the cell through several nuclear pores. The chromatin consists of deoxyribonucleic acid (DNA) and histones that are packaged into chromosomes during mitosis. The nucleolus, which is the dense central portion of the nucleus, manufactures ribosomes. Functions of the nucleus include the storage of genetic material, production of ribosomes, and transcription of ribonucleic acid (RNA).

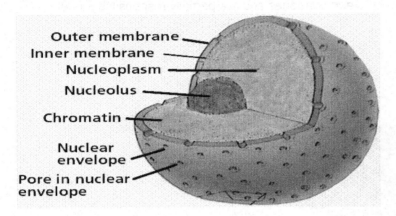

Chloroplasts

Chloroplasts are large organelles called plastids that are enclosed in a double membrane. Discs called thylakoids are arranged in stacks called grana (singular, granum). The thylakoids have chlorophyll molecules on their surfaces. Stromal lamellae separate the thylakoid stacks. Sugars are formed in the stroma, which is the inner portion of the chloroplast. Chloroplasts perform photosynthesis and make food in the form of sugars for the plant. The light reaction stage of photosynthesis occurs in the grana, and the dark reaction stage of photosynthesis occurs in the stroma. Chloroplasts have their own DNA and can reproduce by fission independently.

Chloroplast

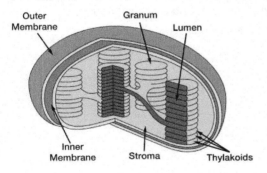

Mitochondria

Mitochondria break down sugar molecules and produce energy in the form of molecules of adenosine triphosphate (ATP). Plant and animal cells contain mitochondria. Mitochondria are enclosed in a bilayer semimembrane of phospholipids and proteins. The intermembrane space is the space between the two layers. The outer membrane has proteins called porins, which allow small molecules through. The inner membrane contains proteins that aid in the synthesis of ATP. The matrix consists of enzymes that help synthesize ATP. Mitochondria have their own DNA and can reproduce by fission independently. Mitochondria also help to maintain calcium concentrations, form blood components and hormones and are partially responsible for cell death.

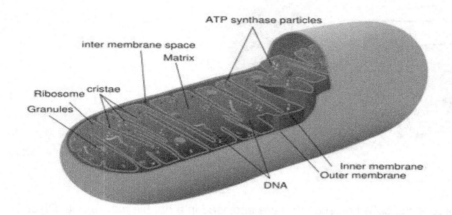

Ribosomes

A ribosome consists of RNA and proteins. Ribosomes are protein factories. Ribosomes translate the code of DNA into proteins by assembling long chains of amino acids. Messenger RNA, mRNA, transports a working copy of the DNA to the ribosome. Transfer RNA, tRNA, collects the needed amino acids and delivers them to the ribosome. The ribosome, which is ribosomal RNA (rRNA), assembles the amino acids into the needed protein. Few ribosomes are free in the cell. Most of the ribosomes in the cell are embedded in the rough endoplasmic reticulum located near the nucleus. Ribosomes consist of two subunits, a large subunit and a small subunit.

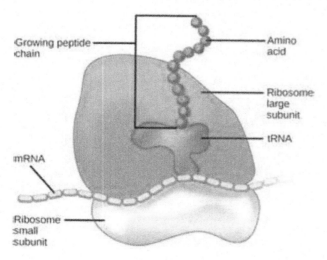

Golgi apparatus

The Golgi apparatus, also called the Golgi body or Golgi complex, is a stack of flattened membranes called *cisternae* that packages, ships, and distributes macromolecules in shipping containers called vesicles. Most Golgi apparatuses have six to eight cisternae. Each Golgi apparatus has four regions: the cis region, the endo region, the medial region, and the trans region. Transfer vesicles from the rough endoplasmic reticulum (ER) enter at the cis region, and secretory vesicles leave the Golgi apparatus from the trans region. The Golgi apparatus directs the movement of carbohydrates, proteins, and lipids throughout the cell. Also, the Golgi apparatus helps modify proteins and lipids before they are shipped.

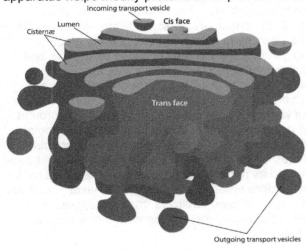

Cytoskeleton

The cytoskeleton is a scaffolding system located in the cytoplasm. The cytoskeleton consists of elongated organelles made of proteins called microtubules, microfilaments, or actin filaments and intermediate filaments. Microtubules consist of polymers of tubulin and microfilaments are composed of actin proteins, while intermediate filaments may consist of any of several different proteins. These organelles provide the shape and the needed support for the cell. They can also give cells the ability to move. These structures assist in moving the chromosomes during mitosis. Microtubules and microfilaments help transport materials throughout the cell. Microtubules are also the major components in cilia and flagella. Microtubules have an average diameter of 25 nanometers.

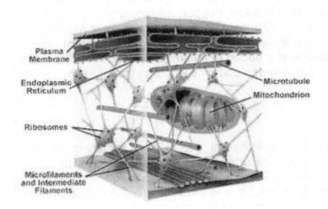

Pollution

Pollution greatly affects the environment, but pollution mitigation has greatly reduced pollution and its effects during the past 40 years. New pollution-control technology has greatly reduced the amount of pollution from new power plants and factories. The Clean Air Act has greatly reduced the amount of pollution. With this legislation, new industrial sites are designed and built with pollution-control technology fully integrated into the facility. These technologies avoid or minimize the negative effects on the environment. New coal-fired power plants are fitted with pollution-control devices that greatly reduce and nearly eliminate sulfur dioxide and nitrogen oxide emissions. This greatly reduces acid rain, improves water quality, and improves the overall health of ecosystems. Reducing acid rain improves soil quality, which in turn improves the health of producers, which consequently improves the health of consumers, essentially strengthening the entire ecosystem. Reduced greenhouse gas emissions have lessened the impact of global warming such as rising sea levels due to melting glaciers and the resulting loss of habitats and biodiversity. Reduced smog and haze improves the intensity of sunlight required for photosynthesis. Nonpoint-source pollution is the leading cause of water pollution in the United States. Nonpoint-source pollution is carried by rainfall or melting snow and is then deposited in lakes, rivers, coastal waters, and other bodies of water. Pollutants can also contaminate ground water and drinking water. Most nonpoint-source pollution is due to agricultural runoff. Nonpoint-source pollution is pollution that does not flow through a pipe, channel, or container. Nonpoint-source pollution comes from many different sources and is difficult to control and regulate. Urban runoff from lawns, streets, and parking lots is treated as nonpoint-source pollution because much of that storm water does not go into a storm drain before entering streams, rivers, lakes, or other bodies of water. Urban runoff contains chemicals such as lawn fertilizers, motor oils, grease, pesticides, soaps, and detergents, each of which is harmful to the environment. For example, detergents harm the protective mucus layer on fish.

Resource management

Resource management such as waste management and recycling greatly impacts the environment. Waste management is the monitoring, collection, transportation, and recycling of waste products. Methods of waste disposal include landfills and incineration. Well-managed landfills include burying wastes, but using clay or another lining material to prevent liquid leachate and layers of soil on top to reduce odors and vermin. Methane gas produced during decomposition can be extracted and burned to generate electricity. Landfills can be used as temporary storage for recyclable materials before transportation to a recycling plant. Wastes can be incinerated to reduce waste volume. Hazardous biomedical waste can be incinerated. However, incineration does emit pollutants and greenhouse gases. Proper waste management always includes recycling. Recycling is a method to recover resources. Recycled materials can be reprocessed into new products. Metals such as aluminum, copper, and steel are recycled. Plastics, glass, and paper products can be recycled. Organic materials such as plant materials and food scraps can be composted. The current trend is to shift from waste management to resource recovery. Wastes should be minimized and reduced to minimize the need for disposal. Unavoidable nonrecyclable wastes should be converted to energy by combustion if at all possible.

Global warming

Global warming caused largely by greenhouse gas emission will greatly affect society in the next several years. Overall global increasing temperatures include increases over land, over the ocean, in the ocean, and even in the troposphere. This increasing temperature leads to more extreme weather events such hurricanes, tornadoes, floods, and droughts. Rising temperatures mean warmer summers. Warmer temperatures may shift tourism and improve agriculture, but global mortality rates may rise due to hotter heat waves. Rising temperatures cause weather patterns to shift, leading to more floods and droughts. Rising temperatures globally means a decrease in glaciers, sea ice, ice sheets, and snow cover, which all contribute to rising sea levels.

Rising sea levels lead to habitat change or loss, which greatly affects numerous species. Some motile species are already moving north to cooler climates. Earlier snowmelt and runoff may overwhelm water management systems. Diseases such as malaria that are spread by mosquitos could spread further, possibly even to temperate regions. Rising sea levels mean higher storm surges and related issues.

Endangered Species Act

The Endangered Species Act (ESA) of 1973 has had a tremendous positive impact on many species. The need for the act was evident because history had shown the loss of biodiversity through the near-extinction of bison and the whooping crane and the extinction of the passenger pigeon. The law was designed to protect "imperiled species" from extinction due to factors such as loss of habitat, overhunting, and lack of conservation. The ESA also protects the species' ecosystems and removes threats to those ecosystems. If an animal is placed on the endangered or threatened list, it is prohibited to "harass, harm, pursue, hunt, shoot, wound, kill, trap, capture, or collect, or to attempt to engage in any such conduct" with any of those animals. The populations of many species including the whooping crane, the gray wolf, the red wolf, and the Hawaiian goose have increased significantly. Some species have even been removed from the endangered species list including the bald eagle, the peregrine falcon, the gray whale, and the grizzly bear. The Endangered Species Act has protected numerous species while balancing human economic needs and rights to private property. Also, the enforcement of the ESA only rarely interferes with land developments. Usually, creative alternatives are available.

National Park System

The National Park System states that its purpose is to "conserve the scenery and the natural and historic objects and the wildlife therein and to provide for the enjoyment of the same in such manner and by such means as will leave them unimpaired for the enjoyment of future generations." The National Park System protects complete ecosystems and houses great biodiversity. National parks are an integral part of many species survival. National parks provide a home to hundreds of endangered or threatened species. Sadly, biodiversity is threatened even within national parks. Although many are vast, they still may not be large enough to support a species population. Studies show that preserved habitats near national parks helps many species better survive. This will prevent fragmentation and further habitat loss. National parks may be threatened by invasion species or pressure for use of land along park boundaries.

Extraction of mineral and energy resources

The extraction of mineral and energy resources by mining and drilling has harmful effects on the environment including pollution and alterations to ecosystems. Mining requires large amounts of land, which harms habitats and affects biodiversity. Mining causes water pollution. Rainwater mixes with the heavy metals in mines and produces an acid runoff that harms aquatic life, birds, and mammals. The pollution is especially bad in countries without proper mining regulations. Water pollution can contaminate drinking water, causing serious human health risks. Open-pit mines and mountaintop removal techniques are especially harsh to the environment, and reclamation is often not regulated in developing countries. Mining often requires large-scale deforestation leading to a loss of habitat for many species. Toxic chemicals such as mercury and sulfuric acid are used in the mining process and are released into bodies of water, harming the aquatic life. If these toxic chemicals are leaked, the ground water is polluted. Oil drilling is controversial due to habitat disruption or loss. Oil spills are toxic to wildlife and difficult to clean up. Offshore drilling uses seismic waves to locate oil, which disturbs whales and dolphins and has been tied to hundreds of beached whales.

Renewable and sustainable resources

The management of natural resources and the renewability or sustainability of those natural resources greatly impact society. Sustainable agriculture involves growing foods in economical ways that do not harm resources. If left unchecked, farming can deplete the soil of valuable nutrients. Crops grown in these depleted soils are less healthy and more susceptible to disease. Sustainable agriculture uses more effective pest control such as insect-resistant corn, which reduces runoff and water pollution in the surrounding area. Sustainable forestry involves replenishing trees as trees are being harvested, which maintains the environment. Energy sources such as wind, solar power, and biomass energy are all renewable. Wind power is clean with no pollution and no greenhouse gas emissions. Cons of wind power include the use of land for wind farms, threats to birds, and the expense to build. Solar power has no greenhouse gas emissions, but some toxic metal wastes result in the production of photovoltaic cells, and solar power requires large areas of land. Biomass energy is sustainable, but its combustion produces greenhouse emissions. Farming biomass requires large areas of land. Fossil fuels, which are nonrenewable, cause substantially more air pollution and greenhouse gas emissions, contributing to habitat loss and global warming.

Practice Test

1. The hydrogen bonds in a water molecule make water a good
   a. Solvent for lipids
   b. Participant in replacement reactions
   c. Surface for small particles and living organisms to move across
   d. Solvent for polysaccharides such as cellulose
   e. Example of an acid

2. The breakdown of a disaccharide releases energy which is stored as ATP. This is an example of a(n)
   a. Combination reaction
   b. Replacement reaction
   c. Endothermic reaction
   d. Exothermic reaction
   e. Thermodynamic reaction

3. Which of the following metabolic compounds is composed of only carbon, oxygen, and hydrogen?
   a. Phospholipids
   b. Glycogen
   c. Peptides
   d. RNA
   e. Vitamins

4. When an animal takes in more energy that it uses over an extended time, the extra chemical energy is stored as:
   a. Fat
   b. Starch
   c. Protein
   d. Enzymes
   e. Cholesterol

5. Which of the following molecules is thought to have acted as the first enzyme in early life on earth?
   a. Protein
   b. RNA
   c. DNA
   d. Triglycerides
   e. Phospholipids

6. Which of the following organelles is/are formed when the plasma membrane surrounds a particle outside of the cell?
   a. Golgi bodies
   b. Rough endoplasmic reticulum
   c. Lysosomes
   d. Secretory vesicles
   e. Endocytic vesicles

7. Which of the following plant organelles contain(s) pigment that give leaves their color?
   a. Centrioles
   b. Cell walls
   c. Chloroplasts
   d. Central vacuole
   e. Golgi apparatus

8. All but which of the following processes are ways of moving solutes across a plasma membrane?
    a. Osmosis
    b. Passive transport
    c. Active transport
    d. Facilitated diffusion
    e. Endocytosis

9. Prokaryotic and eukaryotic cells are similar in having which of the following?
    a. Membrane-bound organelles
    b. Protein-studded DNA
    c. Presence of a nucleus
    d. Integral membrane proteins in the plasma membrane
    e. Flagella composed of microtubules

10. Which of the following cell types has a peptidoglycan cell wall?
    a. Algae
    b. Bacteria
    c. Fungi — chitin
    d. Land plants
    e. Protists

11. Enzymes catalyze biochemical reactions by
    a. Lowering the potential energy of the products
    b. Separating inhibitors from products
    c. Forming a complex with the products
    d. Lowering the activation energy of the reaction
    e. Providing energy to the reaction

12. Which of the following is an example of a cofactor?
    a. Zinc
    b. Actin
    c. Cholesterol
    d. GTP
    e. Chlorophyll

13 Cyanide is a poison that binds to the active site of the enzyme cytochrome c and prevents its activity. Cyanide is a(n)
    a. Prosthetic group
    b. Cofactor
    c. Coenzyme
    d. Inhibitor
    e. Reverse regulator

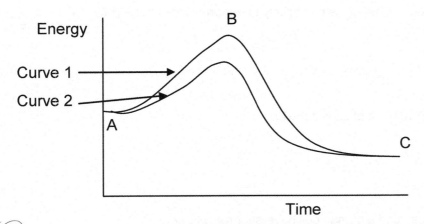

Energy

Curve 1

Curve 2

A

B

C

Time

14. The graph above shows the potential energy of molecules during the process of a chemical reaction. All of the following may be true EXCEPT
   a. This is an endergonic reaction
   b. The activation energy in curve 2 is less than the activation energy in curve 1
   c. The energy of the products is less than the energy of the substrate
   d. Curve 2 shows the reaction in the presence of an enzyme
   e. The reaction required ATP

15. Which of the following is not a characteristic of enzymes?
   a. They change shape when they bind their substrates
   b. They can catalyze reactions in both forward and reverse directions
   c. Their activity is sensitive to changes in temperature
   d. They are always active on more than one kind of substrate
   e. They may have more than one binding site

16. In a strenuously exercising muscle, NADH begins to accumulate in high concentration. Which of the following metabolic process will be activated to reduce the concentration of NADH?
   a. Glycolysis
   b. The Krebs cycle
   c. Lactic acid fermentation
   d. Oxidative phosphorylation
   e. Acetyl CoA synthesis

17. Which of the following statements regarding chemiosmosis in mitochondria is not correct?
   a. ATP synthase is powered by protons flowing through membrane channels
   b. Energy from ATP is used to transport protons to the intermembrane space
   c. Energy from the electron transport chain is used to transport protons to the intermembrane space
   d. An electrical gradient and a pH gradient both exist across the inner membrane
   e. The waste product of chemosmosis is water

18. In photosynthesis, high-energy electrons move through electron transport chains to produce ATP and NADPH. Which of the following provides the energy to create high energy electrons?
   a. NADH
   b. NADP+
   c. O2
   d. Water
   e. Light

19. Which of the following kinds of plants is most likely to perform CAM photosynthesis?
    a. Mosses
    b. Grasses
    c. Deciduous trees
    d. Cacti
    e. Legumes

20. The combination of DNA with histones is called
    a. A centromere
    b. Chromatin
    c. A chromatid
    d. Nucleoli
    e. A plasmid

21. How many chromosomes does a human cell have after meiosis I?
    a. 92
    b. 46
    c. 23
    d. 22
    e. 12

22. In plants and animals, genetic variation is introduced during
    a. Crossing over in mitosis
    b. Chromosome segregation in mitosis
    c. Cytokinesis of meiosis
    d. Anaphase I of meiosis
    e. Anaphase II of meiosis

23. DNA replication occurs during which of the following phases?
    a. Prophase I
    b. Prophase II
    c. Interphase I
    d. Interphase II
    e. Telophase I

24. The synaptonemal complex is present in which of the following phases of the cell cycle?
    a. Metaphase of mitosis
    b. Metaphase of meiosis I
    c. Telophase of meiosis I
    d. Metaphase of meiosis II
    e. Telophase of meiosis II

25. A length of DNA coding for a particular protein is called a(n)
    a. Allele
    b. Genome
    c. Gene
    d. Transcript
    e. Codon

26. In DNA replication, which of the following enzymes is required for separating the DNA molecule into two strands?
    a. DNA polymerase
    b. Single strand binding protein
    c. DNA gyrase
    d. Helicase
    e. Primase

27. Which of the following chemical moieties forms the backbone of DNA?
    a. Nitrogenous bases
    b. Glycerol
    c. Amino groups
    d. Pentose and phosphate
    e. Glucose and phosphate

28. Required for the activity of DNA polymerase
    a. Okazaki fragments
    b. RNA primer
    c. Single-strand binding protein
    d. Leading strand
    e. Replication fork

29. Substrate for DNA ligase
    a. Okazaki fragments
    b. RNA primer
    c. Single-strand binding protein
    d. Leading strand
    e. Replication fork

30. Which of the following is true of the enzyme telomerase?
    a. It is active on the leading strand during DNA synthesis
    b. It requires a chromosomal DNA template
    c. It acts in the 3′ → 5′ direction
    d. It adds a repetitive DNA sequence to the end of chromosomes
    e. It takes the place of primase at the ends of chromosomes

31. Which enzyme in DNA replication is a potential source of new mutations?
    a. DNA ligase
    b. Primase
    c. DNA gyrase
    d. DNA polymerase
    e. Topoisomerase

32. Which of the following mutations is most likely to have a dramatic effect on the sequence of a protein?
    a. A point mutation
    b. A missense mutation
    c. A deletion
    d. A silent mutation
    e. A proofreading mutation

33. Which of the following could be an end product of transcription?
    a. rRNA
    b. DNA
    c. Protein
    d. snRNP
    e. Amino acids

34. The *lac* operon controls
    a. Conjugation between bacteria
    b. Chromatin organization
    c. Gene transcription
    d. Excision repair
    e. Termination of translation

35. All of the following are examples ways of controlling eukaryotic gene expression EXCEPT
    a. Regulatory proteins
    b. Nucleosome packing
    c. Methylation of DNA
    d. RNA interference
    e. Operons

36. Transfer of DNA between bacteria using a narrow tube called a pilus is called
    a. Transformation
    b. Transduction
    c. Operation
    d. Conjugation
    e. Conformation

37. A virus that has incorporated into the DNA of its host
    a. Lysogenic cycle
    b. Lytic cycle
    c. Retrovirus
    d. Provirus
    e. Bacteriophage

38. A virus in this stage is actively replicating DNA
    a. Lysogenic cycle
    b. Lytic cycle
    c. Retrovirus
    d. Provirus
    e. Bacteriophage

39. A bacterial mini-chromosome used in recombinant DNA technology is called a
    a. Centromere
    b. Telomere
    c. Plasmid
    d. Transposon
    e. cDNA

40. Which of the following parts of an angiosperm give rise to the fruit?
    a. Pedicel
    b. Filament
    c. Sepal
    d. Ovary
    e. Meristem

41. Which of the following structures is NOT present in gymnosperms?
    a. Leaves
    b. Pollen
    c. Flowers
    d. Stomata
    e. Roots

42. Which of the following plant structures allows for gas exchange?
    a. Xylem
    b. Phloem
    c. Cuticle
    d. Meristem
    e. Stomata

43. Which type of plant has leaves with parallel veins?
    a. Monocots
    b. Dicots
    c. Angiosperms
    d. Gymnosperms
    e. Nonvascular plants

44. Which type of plant does not produce fruits
    a. Monocots
    b. Dicots
    c. Angiosperms
    d. Gymnosperms
    e. Nonvascular plants

45. Which type of plant produces seeds that are housed inside a fruit
    a. Monocots
    b. Dicots
    c. Angiosperms
    d. Gymnosperms
    e. Nonvascular plants

Questions 46 and 47 pertains to the following diagram representing a cross section of a tree trunk

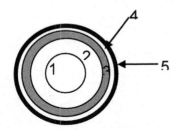

46. Which structure contains tissue that is dead at maturity?
    a. 1
    b. 2
    c. 3
    d. 4
    e. 5

47. Which structure transports carbohydrates to the roots?
    a. 1
    b. 2
    c. 3
    d. 4
    e. 5

48. In ferns, the joining of egg and sperm produces a zygote, which will grow into the
    a. Gametophyte
    b. Sporophyte
    c. Spore
    d. Sporangium
    e. Seedling

49. Which of the following is an example of the alternation of generations life cycle?
    a. Asexual reproduction of strawberries by runners
    b. Annual plants that live through a single growing season
    c. Ferns that have a large diploid and a diminutive haploid stage
    d. Insects that have distinct larval and adult stages
    e. Reptiles that have long periods of dormancy and metabolic inactivity

Questions 50 and 51 pertains to the following diagram of a complete, perfect flower

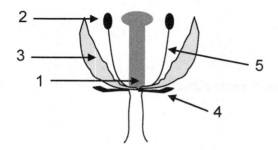

50. The structure in which microspores are produced.
    a. 1
    b. 2
    c. 3
    d. 4
    e. 5

51. The structures composed solely of diploid cells
    a. 1, 2, and 3
    b. 2, 3, and 4
    c. 3, 4, and 5
    d. 1, 4, and 5
    e. 1, 2, and 4

52. Auxins are plant hormones that are involved in all but which of the following processes?
    a. Fruit ripening
    b. Gravitropism
    c. Growth
    d. Phototropism
    e. Seed germination

53. Which of the following plant hormones is most likely to delay aging when sprayed on cut flowers and fruit?
    a. Ethylene
    b. Gibberellins
    c. Cytokinins
    d. Abscisic acid
    e. Jasmonic acid

54. Which of the following would most likely be disruptive to the flowering time of a day-neutral plant?
    a. Daylight interrupted by a brief dark period
    b. Daylight interrupted by a long dark period
    c. High daytime temperatures
    d. Night interrupted by a brief exposure to red light
    e. Night interrupted by a long exposure to red light

55. Animals exchange gases with the environment in all of the following ways EXCEPT
    a. Direct exchange through the skin
    b. Exchange through gills
    c. Stomata
    d. Tracheae
    e. Lungs

56. Which of the following blood components is involved in blood clotting?
    a. Red blood cells
    b. Platelets
    c. White blood cells
    d. Leukocytes
    e. Plasma

57. Which section of the digestive system is responsible for water reabsorption?
    a. The large intestine
    b. The duodenum
    c. The small intestine
    d. The gallbladder
    e. The stomach

58. When Ca^{2+} channels open in a presynaptic cell (doesn't the cell also depolarize?)
    a. The cell depolarizes
    b. The cell hyperpolarizes
    c. An action potential is propagated
    d. Synaptic vesicles release neurotransmitter
    e. The nerve signal is propagated by salutatory conduction

59. Which of the following processes is an example of positive feedback?
    a. High CO2 blood levels stimulate respiration which decreases blood CO2 levels
    b. High blood glucose levels stimulate insulin release, which makes muscle and liver cells take in glucose
    c. Increased nursing stimulates increased milk production in mammary glands
    d. Low blood oxygen levels stimulate erythropoietin production which increases red blood cell production by bone marrow
    e. Low blood calcium levels stimulate parathyroid hormone release from the parathyroid gland. Parathyroid hormone stimulates calcium release from bones.

60. Which of the following would be the most likely means of thermoregulation for a mammal in a cold environment?
    a. Adjusting body surface area
    b. Sweating
    c. Countercurrent exchange
    d. Muscle contractions
    e. Increased blood flow to extremities

61. Which hormone is *not* secreted by a gland in the brain?
    a. Human chorionic gonadotropin (HCG)
    b. Gonadotropin releasing hormone (GnRH)
    c. Luteinizing hormone (LH)
    d. Follicle stimulating hormone (FSH)
    e. None of these

62. Which hormone is secreted by the placenta throughout pregnancy?
    a. Human chorionic gonadotropin (HCG)
    b. Gonadotropin releasing hormone (GnRH)
    c. Luteinizing hormone (LH)
    d. Follicle stimulating hormone (FSH)
    e. None of these

63. Polar bodies are a by-product of
    a. Meiosis I
    b. Meiosis II
    c. Both meiosis I and II
    d. Zygote formation
    e. Mitosis of the morula

64. Which of the following hormones triggers ovulation in females?
    a. Estrogen
    b. Progesterone
    c. Serotonin
    d. Luteinizing hormone
    e. Testosterone

65. Spermatogenesis occurs in the
    a. Prostate gland
    b. Vas deferens
    c. Seminal vesicles
    d. Penis
    e. Seminiferous tubules

66. In which of the following stages of embryo development are the three primary germ layers first present?
    a. Zygote
    b. Gastrula
    c. Morula
    c. Blastula
    e. Coelomate

67. Which of the following extraembryonic membranes is an important source of nutrition in many non-human animal species but NOT in humans?
    a. Amnion
    b. Allantois
    c. Yolk sac
    d. Chorion
    e. Placenta

68. Which of the following is not a mechanism that contributes to cell differentiation and development in embryos?
    a. Asymmetrical cell division
    b. Asymmetrical cytoplasm distribution
    c. Organizer cells
    d. Location of cells on the lineage map
    e. Homeotic genes

69. Which of the following is true of the gastrula?
    a. It is a solid ball of cells
    b. It has three germ layers
    c. It is an extraembryonic membrane
    d. It gives rise to the blastula
    e. It derives from the zona pellucida

70. In birds, gastrulation occurs along the
    a. Dorsal lip of the embryo
    b. Embryonic disc
    c. Primitive streak
    d. Circular blastopore
    e. Inner cell mass

71. In snapdragons, the red (R) allele is incompletely dominant to the white (r) allele. If you saw a pink snapdragon, you would know
    a. Its phenotypes for both parents
    b. Its genotypes for both parents
    c. Its genotype for one parent
    d. Its genotype
    e. Its phenotype but not its genotype

72. In peas, purple flower color (P) is dominant to white (p) and tall stature (T) is dominant to dwarf (t). If the genes are unlinked, how many tall plants will be purple in the progeny of a PpTt x PpTT cross?
    a. 0
    b. ¼
    c. ½
    d. ¾
    e. 1

73. Which of the following does not obey the law of independent assortment?
    a. Two genes on opposite ends of a chromosome
    b. Flower color and height in snapdragons
    c. Two genes on separate chromosomes
    d. Seed color and flower color in peas
    e. Two genes next to each other on a chromosome

74. In a dihybrid cross between bean plants with red (R) wrinkled (w) seeds and white (r) smooth (W) seeds, the F1 progeny is all red and smooth. If the F1 plants are selfed, what proportion of the F2 will also be red and smooth if the genes are linked?
    a. All of them
    b. ¼
    c. 1/2
    d. 9/16
    e. None of them

75. Red-green color blindness is an X-linked trait. What is the probability that a mother that is heterozygous for this trait and a father with this trait will have affected children?
    a. 0
    b. ¼
    c. ½
    d. ¾
    e. 1

76. An individual with an AB blood type needs a blood transfusion. Which of the following types could NOT be a donor?
    a. O
    b. AB
    c. A
    d. B
    e. All can be donors

77. In humans, more than one gene contributes to the trait of hair color. This is an example of
    a. Pleiotropy
    b. Polygenic inheritance
    c. Codominance
    d. Linkage
    e. Epistasis

78. A child is born with type A blood and his mother has type A. Which of the following is NOT a possible combination of genotypes for the mother and father?
    a. IAIB and ii
    b. IAi and ii
    c. IA i and IB i
    d. IAi and IBIB
    e. IAIB and IBi

79. On a standard biomass pyramid, level 3 corresponds to which trophic level?
    a. Producers
    b. Decomposers
    c. Primary consumers
    d. Primary carnivores
    e. Secondary carnivores

    grass → cow → wolf → vulture

80. In the food chain above, vultures represent
    a. Scavengers
    b. Detritivores
    c. Primary carnivores
    d. Herbivores
    e. Secondary consumers

81. Which of the following is the major way in which carbon is released into the environment?
    a. Transpiration
    b. Respiration
    c. Fixation
    d. Sedimentation
    e. Absorption

82. What is the largest reservoir of nitrogen on the planet?
   a. The ocean
   b. Plants
   c. Soil
   d. The atmosphere
   e. Sediments, including fossil fuels

83. The diagram below represents the three types of survivorship curves, describing how mortality varies as species age. Which of the following species is most likely to exhibit Type I survivorship?

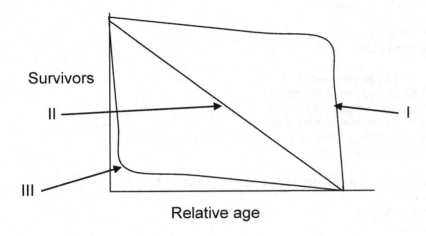

   a. Frogs
   b. Oysters
   c. Salmon
   d. Dolphins
   e. Shrimp

84. A population of 1000 individuals has 110 births and 10 deaths in a year. Its growth rate (r) is equal to
   a. 0.01 per year
   b. 0.1 per year
   c. 0.09 per year
   d. 0.11 per year
   e. 0.009 per year

85. During primary succession, which species would most likely be a pioneer species?
   a. Lichens
   b. Fir trees
   c. Mosquitoes
   d. Dragonflies
   e. Mushrooms

86. Which of the following habitats would provide an opportunity for secondary succession?
   a. A retreating glacier
   b. Burned cropland
   c. A newly formed volcanic island
   d. A 500 year old forest
   e. A sand dune

87. Which biome is most likely to support the growth of epiphytes?
   a. Deserts
   b. Tropical rain forests
   c. Temperate deciduous forests
   d. Taigas
   e. Savannas

88. Which of the following is NOT a natural dispersal process that would lead to species colonization on an island?
   a. Mussels carried into a lake on the hull of a ship
   b. Drought connecting an island to other land
   c. Floating seeds
   d. Animals swimming long distances
   e. Birds adapted to flying long distances

89. When a population reaches its carrying capacity
   a. Other populations will be forced out of the habitat
   b. Density-dependent factors no longer play a role
   c. Density-independent factors no longer play a role
   d. The population growth rate approaches zero
   e. The population size begins to decrease

90. Which of the following is an example of a density-dependent limiting factor?
   a. Air pollution by a factory
   b. The toxic effect of waste products
   c. Nearby volcanic eruptions
   d. Frosts
   e. Fires

91. Two species of finches are able to utilize the same food supply, but their beaks are different. They are able to coexist on an island because of
   a. Niche overlap
   b. Character displacement
   c. Resource partitioning
   d. Competitive exclusion
   e. Realized niches

92. Lichens consist of fungi and algae. The algae supply sugars through performing photosynthesis while the fungi provide minerals and a place to attach. This is an example of
   a. Mutualism
   b. Commensalism
   c. Parasitism
   d. Coevolution
   e. Resource partitioning

93. Which of the following of Lamarck's evolutionary ideas turned out to be true?
   a. Natural selection
   b. Organisms naturally transform into increasingly complex organisms
   c. Inheritance of acquired characters
   d. Body parts develop with increased usage and weaken with disuse
   e. Genes are the basic units of inheritance

94. The weight of adult wolves is within a fairly narrow range, even if they are well-fed in zoos. This is an example of
    a. Stabilizing selection
    b. Directional selection
    c. Disruptive selection
    d. Sexual selection
    e. Artificial selection

95. Which of the following is a trait that results from disruptive selection?
    a. Insecticide resistance
    b. Male peacocks have colorful plumage while females do not
    c. Within the same species, some birds have large bills, while others have small bills.
    d. Human height
    e. Various varieties of wheat

96. Which of the following conditions would promote evolutionary change?
    a. Neutral selection
    b. Random mating
    c. A large population
    d. An isolated population
    e. Gene flow

97. Which of the following would create the greatest amount of genetic variation for a diploid species in a single generation?
    a. Crossing over
    b. Mutation
    c. Hybridization
    d. Independent assortment of homologs
    e. Random joining of gametes

98. A population of pea plants has 25% dwarf plants and 75% tall. The tall allele, T is dominant to dwarf (t). What is the frequency of the T allele?
    a. 0.75
    b. 0.67
    c. 0.5
    d. 0.25
    e. 0.16

99. Darwin's idea that evolution occurs by the gradual accumulation of small changes can be summarized as
    a. Punctuated equilibrium
    b. Phyletic gradualism
    c. Convergent evolution
    d. Adaptive radiation
    e. Sympatric speciation

100. Which of the following processes of speciation would most likely occur if a species of bird were introduced into a group of islands that were previously uninhabited by animals?
    a. Allopatric speciation
    b. Adaptive radiation
    c. Sympatric speciation
    d. Artificial speciation
    e. Hybridizing speciation

Answers and Explanations

1. C: The hydrogen bonds between water molecules cause water molecules to attract each other (negative pole to positive pole. and "stick" together. This gives water a high surface tension, which allows small living organisms, such as water striders, to move across its surface. Since water is a polar molecule, it readily dissolves other polar and ionic molecules such as carbohydrates and amino acids. Polarity alone is not sufficient to make something soluble in water, however; for example, cellulose is polar but its molecular weight is so large that it is not soluble in water.

2. D: An exothermic reaction releases energy, whereas an endothermic reaction requires energy. The breakdown of a chemical compound is an example of a decomposition reaction ($AB \rightarrow A + B$.. A combination reaction ($A + B \rightarrow AB$. is the reverse of a decomposition reaction, and a replacement (displacement) reaction is one where compound breaks apart and forms a new compound plus a free reactant ($AB + C \rightarrow AC + B$ or $AB + CD \rightarrow AD + CB$.

3. B: Glycogen is a polysaccharide, a molecule composed of many bonded glucose molecules. Glucose is a carbohydrate, and all carbohydrates are composed of only carbon, oxygen, and hydrogen. Most other metabolic compounds contain other atoms, particularly nitrogen, phosphorous, and sulfur.

4. A: Long term energy storage in animals takes the form of fat. Animals also store energy as glycogen, and plants store energy as starch. , but these substances are for shorter-term use. Fats are a good storage form for chemical energy because fatty acids bond to glycerol in a condensation reaction to form fats (triglycerides). This reaction, which releases water, allows for the compacting of high-energy fatty acids in a concentrated form.

5. B: Some RNA molecules in extant organisms have enzymatic activity; for example the formation of peptide bonds on ribosomes is catalyzed by an RNA molecule. This and other information has led scientists to believe that the most likely molecules to first demonstrate enzymatic activity were RNA molecules.

6. E: Endocytosis is a process by which cells absorb larger molecules or even tiny organisms, such as bacteria, than would be able to pass through the plasma membrane. Endocytic vesicles containing molecules from the extracellular environment often undergo further processing once they enter the cell.

7. C: Chloroplasts contain the light-absorbing compound chlorophyll, which is essential in photosynthesis. This gives leaves their green color. Chloroplasts also contain yellow and red carotenoid pigments, which give leaves red and yellow colors in the fall as chloroplasts lose their chlorophyll.

8. A: Osmosis is the movement of water molecules (not solutes) across a semi-permeable membrane. Water moves from a region of higher concentration to a region of lower concentration. Osmosis occurs when the concentrations of a solute differ on either side of a semi-permeable membrane. For example, a cell (containing a higher concentration of water) in a salty solution (containing a lower concentration of water) will lose water as water leaves the cell. This continues until the solution outside the cell has the same salt concentration as the cytoplasm.

9. D: Both prokaryotes and eukaryotes interact with the extracellular environment and use membrane-bound or membrane-associated proteins to achieve this. They both use diffusion and active transport to move materials in and out of their cells. Prokaryotes have very few proteins associated with their DNA, whereas eukaryotes' DNA is richly studded with proteins. Both types of living things can have flagella, although with different structural characteristics in the two groups. The most important differences between prokaryotes and eukaryotes are the lack of a nucleus and membrane-bound organelles in prokaryotes.

10. B: Bacteria and cyanobacteria have cell walls constructed from peptidoglycans – a polysaccharide and protein molecule. Other types of organisms with cell walls, for instance, plants and fungi, have cell walls composed of different polysaccharides. Plant cell walls are composed of cellulose, and fungal cell walls are composed of chitin.

11. D: Enzymes act as catalysts for biochemical reactions. A catalyst is not consumed in a reaction, but, rather, lowers the activation energy for that reaction. The potential energy of the substrate and the product remain the same, but the activation energy—the energy needed to make the reaction progress—can be lowered with the help of an enzyme.

12. A: A cofactor is an inorganic substance that is required for an enzymatic reaction to occur. Cofactors bind to the active site of the enzyme and enable the substrate to fit properly. Many cofactors are metal ions, such as zinc, iron, and copper.

13. D: Enzyme inhibitors attach to an enzyme and block substrates from entering the active site, thereby preventing enzyme activity. As stated in the question, cyanide is a poison that irreversibly binds to an enzyme and blocks its active site, thus fitting the definition of an enzyme inhibitor.

14. A: Because the energy of the products is less than the energy of the substrate, the reaction releases energy and is an exergonic reaction.

15. D: Enzymes are substrate-specific. Most enzymes catalyze only one biochemical reaction. Their active sites are specific for a certain type of substrate and do not bind to other substrates and catalyze other reactions.

16. C: Lactic acid fermentation converts pyruvate into lactate using high-energy electrons from NADH. This process allows ATP production to continue in anaerobic conditions by providing NAD^+ so that ATP can be made in glycolysis.

17. B: Proteins in the inner membrane of the mitochondrion accept high-energy electrons from NAD and $FADH_2$, and in turn transport protons from the matrix to the intermembrane space. The high proton concentration in the intermembrane space creates a gradient which is harnessed by ATP synthase to produce ATP.

18. E: Electrons trapped by the chlorophyll P680 molecule in photosystem II are energized by light. They are then transferred to electron acceptors in an electron transport chain.

19. D: CAM photosynthesis occurs in plants that grow where water loss must be minimized, such as cacti. These plants open their stomata and fix CO_2 at night. During the day, stomata are closed, reducing water loss. Thus, photosynthesis can proceed without water loss.

20. B: DNA wrapped around histone proteins is called chromatin. In a eukaryotic cell, DNA is always associated with protein; it is not "naked" as with prokaryotic cells.

21. B: The diploid chromosome number for humans is 46. After DNA duplication but before the first cell division of meiosis, there are 92 chromosomes (46 pairs). After meiosis I is completed, the chromosome number is halved and equals 46. Each daughter cell is haploid, but the chromosomes are still paired (sister chromatids). During meiosis II, the two sister chromatids of each chromosome separate, resulting in 23 haploid chromosomes per germ cell.

22. D: In anaphase I, homologous chromosome pairs segregate randomly into daughter cells. This means that each daughter cell contains a unique combination of chromosomes that is different from the mother cell and different from its cognate daughter cell.

23. C: Although there are two cell divisions in meiosis, DNA replication occurs only once. It occurs in interphase I, before M phase begins.

24. C: The synaptonemal complex is the point of contact between homologous chromatids. It is formed when nonsister chromatids exchange genetic material through crossing over. Once meiosis I has completed, crossovers have resolved and the synaptonemal complex no longer exists. Rather, sister chromatids are held together at their centromeres prior to separation in anaphase II.

25. C: Genes code for proteins, and genes are discrete lengths of DNA on chromosomes. An allele is a variant of a gene (different DNA sequence.. In diploid organisms, there may be two versions of each gene.

26. D: The enzyme helicase unwinds DNA. It depends on several other proteins to make the unwinding run smoothly, however. Single-strand binding protein holds the single stranded DNA in place, and topoisomerase helps relieve tension at the replication fork.

27. D: DNA is composed of nucleotides joined together in long chains. Nucleotides are composed of a pentose sugar, a phosphate group, and a nitrogenous base. The bases form the "rungs" of the ladder at the core of the DNA helix and the pentose-phosphates are on its outside, or backbone.

28. B: DNA replication begins with a short segment of RNA (not DNA.. DNA polymerase cannot begin adding nucleotides without an existing piece of DNA (a primer).

29. A: DNA synthesis on the lagging strand forms short segments called Okazaki fragments. Because DNA polymerase can only add nucleotides in the 5′ → 3′ direction, lagging strand synthesis is discontinuous. The final product is formed when DNA ligase joins Okazaki fragments together.

30. D: Each time a cell divides; a few base pairs of DNA at the end of each chromosome are lost. Telomerase is an enzyme that uses a built-in template to add a short sequence of DNA over and over at the end of chromosomes—a sort of protective "cap". This prevents the loss of genetic material with each round of DNA replication.

31. D: DNA polymerase does not match base pairs with 100% fidelity. Some level of mismatching is present for all DNA polymerases, and this is a source of mutation in nature. Cells have mechanisms of correcting base pair mismatches, but they do not fix all of them.

32. C: Insertions and deletions cause frameshift mutations. These mutations cause all subsequent nucleotides to be displaced by one position, and thereby cause all the amino acids to be different than they would have been if the mutation had not occurred.

33. A: Transcription is the process of creating an RNA strand from a DNA template. All forms of RNA, for example mRNA, tRNA, and rRNA, are products of transcription.

34. C: The *lac* operon controls transcription of the gene that allows bacteria to metabolize lactose. It codes for both structural and regulatory proteins and includes promoter and operator sequences.

35. E: Operons are common to prokaryotes. They are units of DNA that control the transcription of DNA and code for their own regulatory proteins as well as structural proteins.

36. D: Conjugation is direct transfer of plasmid DNA between bacteria through a pilus. The F plasmid contains genes that enable bacteria to produce pili and is often the DNA that is transferred between bacteria.

37. D: In the lysogenic cycle, viral DNA gets incorporated into the DNA of the host. A virus in this dormant stage is called a provirus. Eventually, an external cue may trigger the virus to excise itself and begin the lytic cycle.

38. B: In the lytic cycle, viruses use host resources to produce viral DNA and proteins in order to create new viruses. They destroy the host cell in the process by lysing it. For this reason, actively replicating viruses are said to be in the lytic cycle.

39. C: Plasmids are small circular pieces of DNA found in bacteria that are widely used in recombinant DNA technology. They are cut with restriction enzymes and DNA of interest is ligated to them. They can then easily be used to transform bacteria.

40. D: The ovary houses the ovules in a flower. Pollen grains fertilize ovules to create seeds, and the ovary matures into a fruit.

41. C: Gymnosperms reproduce by producing pollen and ovules, but they do not have flowers. Instead, their reproductive structures are cones or cone-like structures.

42. E: Stomata are openings on leaves that allow for gas exchange, which is essential for photosynthesis. Stomata are formed by guard cells, which open and close based on their turgidity.

43. A: Monocots differ from dicots in that they have one cotyledon, or embryonic leaf in their embryos. They also have parallel veination, fibrous roots, petals in multiples of three, and a random arrangement of vascular bundles in their stems.

44. E: Nonvascular plants do not produce fruits like angiosperms and gymnosperms do. They generally reproduce sexually, but produce spores instead of seeds.

45. C: Angiosperms produce flowers, with ovules inside of ovaries. The ovaries become a fruit, with seeds inside. Gymnosperms have naked seeds that are produced in cones or cone like structures.

46. A: The actual wood of a tree trunk is made of dead xylem tissue. It does not function in the transport of water, but rather functions only in support.

47. C: The phloem transports carbohydrates from the shoot to the roots. Phloem tissue is living and is located outside the xylem.

48. B: In ferns, the mature diploid plant is called a sporophyte. Sporophytes undergo meiosis to produce spores, which develop into gametophytes, which produce gametes.

49. C: Alternation of generations means the alternation between the diploid and haploid phases in plants.

50. B: Anthers produce microspores (the male gametophytes of flowering plants), which undergo meiosis to produce pollen grains.

51. C: In flowering plants, the anthers house the male gametophytes (which produce sperm) and the pistils house the female gametophytes (which produce eggs). Eggs and sperm are haploid. All other tissues are solely diploid.

52. A: The plant hormone ethylene is responsible for fruit ripening. Auxins are involved in a range of processes involving growth and development.

53. C: Cytokinins stimulate cell division (cytokinesis) and have been found to delay senescence (aging). They are often sprayed on cut flowers and fruit to prolong their shelf life.

54. C: Day-neutral plants are not affected by day length in their flowering times. Rather, they respond to other environmental cues like temperature and water.

55. C: Plants exchange gases with the environment through pores in their leaves called stomata. Animals exchange gases with the environment in many different ways: small animals like flatworms exchange gases through their skin; insects use tracheae; and many species use lungs.

56. B: Platelets are cell fragments that are involved in blood clotting. Platelets are the site for the blood coagulation cascade. Its final steps are the formation of fibrinogen which, when cleaved, forms fibrin, the "skeleton" of the blood clot.

57. A: The large intestine's main function is the reabsorption of water into the body to form solid waste. It also allows for the absorption of vitamin K produced by microbes living inside the large intestine.

58. D: When Ca^{2+} channels open, calcium enters the axon terminal and causes synaptic vesicles to release neurotransmitter into the synaptic cleft.

59. C: In a positive feedback loop, an action intensifies a chain of events that, in turn, intensify the conditions that caused the action beyond normal limits. Nursing stimulates lactation, which promotes nursing. Contractions during childbirth, psychological hysteria, and sexual orgasm are all examples of positive feedback.

60. D: Mammals often warm themselves by altering their metabolism. Shivering warms animals due to the heat generated by contractions in trunk muscles.

61. A: HCG is secreted by the trophoblast, part of the early embryo, following implantation in the uterus. GnRH (gonadotropin-releasing hormone. is secreted by the hypothalamus, while LH (luteinizing hormone. and FSH (follicle-stimulating hormone. are secreted by the pituitary gland. GnRH stimulates the production of LH and FSH. LH stimulates ovulation and the production of estrogen and progesterone by the ovary in females, and testosterone production in males. FSH stimulates maturation of the ovarian follicle and estrogen production in females and sperm production in males.

62. E: The placenta secretes progesterone and estrogen once a pregnancy is established. Early in pregnancy, the placenta secretes hCG.

63. C: In oogenesis, meiosis I produces a secondary oocyte and a polar body. Both the first polar body and the secondary oocyte undergo meiosis II. The secondary oocyte divides to produce the ovum and the second polar body.

64. D: Positive feedback from rising levels of estrogen in the menstrual cycle produces a sudden surge of luteinizing hormone (LH). This high level triggers ovulation.

65. E: The testes contain hundreds of seminiferous tubules for the production of sperm, or spermatogenesis. This requires 64-72 days. Leydig cells surround the seminiferous tubules and produce male sex hormones called androgens, the most important of which is testosterone. Semen is made in the seminal vesicles, prostate gland, and other glands. Sperm are transferred to the penis via the epididymis, where they become motile, and thence through the vas deferens.

66. B: The gastrula is formed from the blastocyst, which contains a bilayered embryonic disc. One layer of this disc's inner cell mass further subdivides into the epiblast and the hypoblast, resulting in the three primary germ layers (endoderm, mesoderm, ectoderm).

67. C: In birds and reptiles, the yolk sac contains the yolk, the main source of nutrients for the embryo. In humans, the yolk sac is empty and embryos receive nutrition through the placenta. However, the yolk sac forms part of the digestive system and is where the earliest blood cells and blood vessels are formed.

68. D: A lineage map describes the fates of cells in the early embryo: in other words, it tells which germ layer different cells will occupy. In some small organisms such as the nematode *Caenorhabditis elegans*, all of the adult cells can be traced back to the egg. A lineage map is not a mechanism of embryo development, but rather a tool for describing it.

69. B: The gastrula is the first three-layered stage of the embryo, containing ectoderm, mesoderm, and endoderm

70. C: In birds, the invagination of gastrulation occurs along a line called a primitive streak. Cells migrate to the primitive streak, and the embryo becomes elongated.

71. D: You would know the snapdragon has an *Rr* genotype, but you would not know whether its parents had an *Rr* genotype or a combination of Rr and *rr* or *RR* and *rr*.

72. D: All the plants will be tall, and flower color will assort independently of stature. In a *Pp* x *Pp* cross, ¾ of the progeny will be purple.

73. E: Two genes next to, or within a specified close distance of, each other, are said to be linked. Linked genes do not follow the law of independent assortment because they are too close together to be segregated from each other in meiosis.

74. C: If the genes are linked, there would be only two kinds of alleles produced by the F1 plants: *Rw* and *rW*. A Punnet square with these alleles reveals that half the progeny will have both an *R* and a *W* allele.

75. C: Half of the boys will receive the color-blind allele from the mother, and the other half will receive the normal one. All the girls will receive the color-blind allele from the father; half of them will also get one from the mother, while the other half will get the normal one. Therefore, half the children will be colorblind.

76. E: An individual with AB blood is tolerant to both the A carbohydrate on red blood cells and the B carbohydrate as "self" and can therefore accept any of the 4 different blood types.

77. B: When more than one gene contributes to a trait, inheritance of that trait is said to be polygenic. This type of inheritance does not follow the rules of Mendelian genetics.

78. D: The parents in D could only have offspring with AB or B blood types, not the A blood type.

79. D: At the lowest trophic level are the producers, followed by primary consumers. Primary carnivores follow consumers, followed by secondary carnivores.

80 A: Vultures eat carrion, or dead animals, so they are considered scavengers. Detritivores are heterotrophs that eat decomposing organic matter such as leaf litter. They are usually small.

81. B: Carbon is released in the form of CO_2 through respiration, burning, and decomposition.

82. D: Most nitrogen is in the atmosphere in the form of N_2. In order for it to be used by living things, it must be fixed by nitrogen-fixing bacteria. These microorganisms convert N_2 to ammonia, which then forms NH_4^+ (ammonium).

83. D: Type I curves describe species in which most individuals survive to middle age, after which deaths increase. Dolphins have few offspring, provide extended care to the young, and live a long time.

84. B: The growth rate is equal to the difference between births and deaths divided by population size.

85. A: Pioneer species colonize vacant habitats, and the first such species in a habitat demonstrate primary succession. Succession on rock or lava often begins with lichens. Lichens need very little organic material and can erode rock into soil to provide a growth substrate for other organisms.

86. B: Secondary succession occurs when a habitat has been entirely or partially disturbed or destroyed by abandonment, burning, storms, etc.

87. B: Epiphytes are plants that grow in the canopy of trees, and the tropical rain forest has a rich canopy because of its density and extensive moisture.

88. A: Transportation by humans or human-associated means is not considered a natural dispersal process.

89. D: Within a habitat, there is a maximum number of individuals that can continue to thrive, known as the habitat's carrying capacity. When the population size approaches this number, population growth will stop.

90. B: Density-dependent limiting factors on population growth are factors that vary with population density. Pollution from a factory, volcanic eruptions, frosts, and fires do not vary as a function of population size. Waste products, however, increase with population density and could limit further population increases.

91. B: Character displacement means that, although similar, species in the same habitat have evolved characteristics that reduce competition between them. It occurs as a result of resource partitioning.

92. A: Because both species benefit, lichens constitute an example of mutualism.

93. D: Natural selection was Darwin's idea, not Lamarck's. Mendel discovered that genes are the basic units of inheritance. Lamarck's observation about use and disuse is true, although he did not connect it with the underlying mechanism of natural selection.

94. A: Stabilizing selection is a form of selection in which a particular trait, such as weight, becomes stable within a population. It results in reduced genetic variability, and the disappearance of alleles for extreme traits. Over time, the most common phenotypes survive.

95. C: Disruptive selection occurs when the environment favors alleles for extreme traits. In the example, seasonal changes can make different types of food available at different times of the year, favoring the large or short bills, respectively.

96. E: Options A-D all describe conditions that would lead to genetic equilibrium, where no evolution would occur. Gene flow, which is the introduction or removal of alleles from a population, would allow natural selection to work and could promote evolutionary change.

97. C: Hybridization between two different species would result in more genetic variation than sexual reproduction within a species.

98. C: According to Hardy-Weinberg equilibrium, $p + q = 1$ and $p^2 + 2pq + q^2 = 1$. In this scenario, $q^2 = 0.25$, so $q = 0.5$. p must also be 0.5.

99. B: Phyletic gradualism is the view that evolution occurs at a more or less constant rate. Contrary to this view, punctuated equilibrium holds that evolutionary history consists of long periods of stasis punctuated by geologically short periods of evolution. This theory predicts that there will be few fossils revealing intermediate stages of evolution, whereas phyletic gradualism views the lack of intermediate-stage fossils as a deficit in the fossil record that will resolve when enough specimens are collected.

100. B: Adaptive radiation is the evolution of several species from a single ancestor. It occurs when a species colonizes a new area and members diverge geographically as they adapt to somewhat different conditions.